Memory Traces in the Brain

Memory Traces
in the Brain

DANIEL L. ALKON, M.D.

Section on Neural Systems,
National Institute of Neurologic
and Communicative Disorders and Stroke
at the Marine Biological Laboratory
Woods Hole, Massachusetts

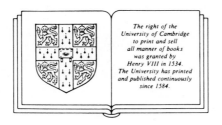

The right of the
University of Cambridge
to print and sell
all manner of books
was granted by
Henry VIII in 1534.
The University has printed
and published continuously
since 1584.

CAMBRIDGE UNIVERSITY PRESS

Cambridge
London New York New Rochelle
Melbourne Sydney

Published by the Press Syndicate of the University of Cambridge
The Pitt Building, Trumpington Street, Cambridge CB2 1RP
32 East 57th Street, New York, NY 10022, USA
10 Stamford Road, Oakleigh, Melbourne 3166, Australia

First published 1987

Printed in the United States of America

Library of Congress Cataloging-in-Publication Data

Alkon, Daniel L.
  Memory traces in the brain.

  1. Memory—Physiological aspects.   I. Title.
QP406.A625   1987   153.1   87-15103
ISBN 0 521 24735 7 hard covers
ISBN 0 521 35867 1 paperback

Contents

Preface

Humans appear, in one important respect, to be no different from other organisms: They seek homeostasis. They strive for equilibrium. This homeostatic orientation is manifest not only by our continual efforts to satisfy obvious biological needs and drives, but by our very human tendency to avoid uncertainty, to seek the familiar. We are comfortable with what we know, disturbed by the new and different, made anxious by the unexpected. We reinforce, attempt to recreate situations, relationships, ideas, and beliefs that we already know from the past. In fact, this is the essence of why learning and memory are so important for understanding and predicting human behavior – we know the past and try to find it in the future by virtue of what we remember.

Avoidance of uncertainty in our system of beliefs is not only an individual process, it is also a collective, a societal phenomenon. Groups to which we belong subscribe to beliefs about themselves, their universe, and pass these beliefs onto their children. Such beliefs must also contribute importantly to our "scientific" knowledge – knowledge that ultimately must be demonstrated and tested and measured.

It is a prime function of science to transcend our need as individuals and groups for certainty – and thereby ultimately arrive at more certainty. If we can suspend our belief systems, we can generate new hypotheses and paradoxically be most ready to abandon these hypotheses in response to observations of phenomena not consistent with our beliefs. If we can live with the anxiety of uncertainty, we can let our minds range free – we can imagine without knowing, we can leap beyond the familiar. If we can live with the anxiety of uncertainty, we can remove the filters from our senses so that we do not discard subtle unanticipated perceptions – so that we can make the most careful, unbiased observations. Then, we can, in fact, learn from our experience and not only as children, but also as mature, even aged adults, modify our views and our behavior.

In the second half of the twentieth century, we arrived at the problems of learning and memory with collective preconceived notions. Mechanisms of learning and memory must occur within our brains, within neural systems of animals. Learning and memory must occur at the junctions between elements of these systems – the synapses. Learning and memory must involve structural changes – growth and formation of new synapses or marked transformations of old neuronal geometry. Memory must become permanent when the neurons make new proteins or different amounts of old proteins.

But must learning and memory occur at the synapse itself or can the crucial information be stored in structures in close proximity to the synapse so that signals crossing the synapse are not changed but are transmitted along proximal structures in a modified form? Must memory involve the synthesis of proteins or are there other biochemical steps that critically determine what information is stored? Do neurons have to actually change their shape, size, and number of junctions with other neurons in order for us to remember? Can we suspend our beliefs to truly ask these questions?

If we can, then we must study the processes themselves. We must study learning and memory as they occur in nature – not as we *think* they occur. Then we must be guided through the jungle of possibilities to the realm of realities, of actualities, by what we can sense, at any moment, in the here and now.

Described here is only one scientific perspective – a picture, sketched and colored of necessity through a haze. This is a picture unavoidably out of focus, relying on the suggested images that emerge from limited bits of information. For me, this picture has an inherent beauty – a beauty, which in its appreciation, offers an opportunity to transcend the very finite concerns of my own survival and well-being. But this picture is not the only one from which I derive intellectual satisfaction, nor is it a picture that has sprung simply from the work of my laboratory or my generation. The beauty of science, luckily, can be shared, and science evolves through this sharing and its transmission from one individual to another. The ultimate sharing of scientific insight provides the global context of all of our individual efforts and shapes our collective understanding.

The focus of this volume on a rather restricted area and mode of inquiry should not convey that its content is in isolation from inquiries of the past or the present. The electrophysiologic techniques and concepts at the center of our studies here were gifts from our predecessors such as Galvani, Bernstein, Cajal, Katz, Hodgkin, Huxley, Cole, and Eccles. The use of "simple" systems to uncover integrative functions of neural networks was pioneered in the work of Hartline, Kuffler, Wiersma, Tauc, Nicholls to name a few. Our notions as to how "complex" vertebrate systems function derive from, among others, von Bekesy, Eccles, Mountcastle, Hubel, Wiesel, Llinas, Ito, and Anderson.

The possibilities of "simple" systems for the exploration of mechanisms of learning were first understood by Horridge, Bruner, Tauc, and Kandel. Kandel and his colleagues later creatively pursued these possibilities. Parallels among associative learning behaviors of animals of vastly different levels of evolution were revealed by von Frisch, Menzel, Davis, Gelperin, and Sahley. Quantitative analyzable features of associative learning were brilliantly derived by Pavlov and later by Thorndike, Hull, Skinner, Gormezano, Rescorla, Wagner, and many more. In my own laboratory, many contributed to the work described here. These include Bank, Collin, Coulter, Crow, Disterhoft, Farley, Goh, Grossman, Harrigan, Heldman, Kuzirian, Lederhendler, Lo Turco, Naito, Neary, Shoukimous, and Tabata. In the text I mention few names, not because I do not recognize the contributions of these many colleagues, past and present. Rather, it is my attempt to present the science itself, observations and hypotheses. It is the science, data and speculation, rather than the personalities that I hope will take center stage.

Finally, I would like to gratefully acknowledge the encouragement, support, critical judgement, and friendship of Harold Atwood, Lynn Bindman, Robert De Lorenzo, Dori Gormezano, Masao Ito, Rodolfo Llinas, John Pfeiffer, Rami Rahaminoff, Howard Rassmussen, Victor Shashoua, Ladislav Tauc, Nakaakira Tsukahara, Charles Woody, and my wife Betty. To find and maintain the courage of one's convictions and equally important the courage to modify those convictions as experience dictates is necessary for any venture into the unknown. On occasion, most of us have experienced the loneliness, albeit a sometimes splendid loneliness, of asking questions passionately, without compromise. Yet, for me, and I suppose for most, that loneliness is tolerable only because of those with whom we do not feel alone. In this sense, as in the historical sense of the evolution of human thought, no quest, no action, ever entirely resides within one individual.

Woods Hole, 1986

1
Introduction

How do we remember? What is it that determines our consciousness and colors the nature of our experience? What has preserved the records of our past and is continually recalled in our present and forever shapes our future? Surely this process of recording and recall is at the core of our very essence as human beings.

When we ask how do we remember, we are really asking several interrelated questions. We are asking first how do we sense stimuli in our environment? How is this sensed or perceived information processed to ultimately result in behavior? How are patterns of sensed and behaviorally expressed information stored? And, ultimately, how is stored information subsequently recalled?

There is now abundant evidence that sensation, integration, and behavioral expression is accomplished by our nervous systems. Groups of neurons linked together by synaptic junctions provide complex pathways along which signals triggered by stimulus patterns travel. The exact nature of these pathways and the integrative tasks they perform are clearly understood in only a relatively few number of organisms. (All the figures in this chapter are to be used for impressions of such networks rather than precise details of the pathways.) Even in relatively simple invertebrate species, neural systems have been comprehensively analyzed usually within limited sections. [Examples include the eye of the horseshoe crab *Limulus*, the segmental ganglia of the medicinal leech *Hirudo*, and the interaction of the visual and vestibular pathway of the nudibranch mollusc *Hermissenda crassicornis* (Figure 1).]

In more complex vertebrate species (such as the rabbit, cat, and monkey), pathways have been determined mainly as relationships between stations of neuron clusters or "nuclei." Touch signals from a limb, for example, are transmitted by sensory cells to cells within the spinal cord. Spinal cord cells can send signals back to muscles to execute behavior (Figure 2), can interact with other spinal cord cells to provide

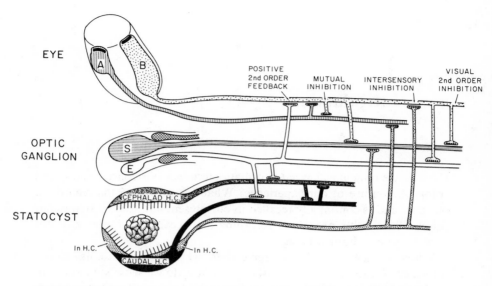

Figure 1. *Hermissenda* neural system (schematic and partial diagram) responsive to light and rotation. Each eye has two type A and three type B photoreceptors; each optic ganglion has 13 second-order visual neurons; each statocyst has 12 hair cells. The neural interactions (intersection of vertical and horizontal processes) identified to be reproducible from preparation-to-preparation are based on intracellular recordings from hundreds of pre- and postsynaptic neuron pairs. In HC, hair cell ∼ 45° lateral to the caudal north–south equatorial pole of statocyst; S, silent optic ganglion cell, electrically coupled to the E cell; E, optic ganglion cell, presynaptic source of EPSPs in type B photoreceptors. The E second-order visual neuron causes EPSPs in type B photoreceptors and cephalad hair cells and simultaneous inhibitory postsynaptic potentials (IPSPs) in caudal hair cells. (From Alkon, 1980)

local integrative functions, or can relay information to nuclei within the brain, which in turn relay information to still other nuclei for further processing (Figure 3). We also know many details about the organization of functional units within particular brain regions. For example, within the vertebrate retina the sensory cells, called rods and cones, send signals to other cells, called bipolar and horizontal cells, which in turn communicate with amacrine and ganglion cells (Figure 4), which finally relay information to more central neural clusters or nuclei of the brain (including the lateral geniculate, superior colliculus) and, ultimately, the visual cortex (Figure 5). Similar knowledge has been accumulated about functional units within the hippocampus and the cerebellum (Figures 6 and 7). Despite this large body of accumulated knowledge of neural organization, however, we are still very far from being able to precisely describe exactly how signals flow along discrete neuronal pathways to

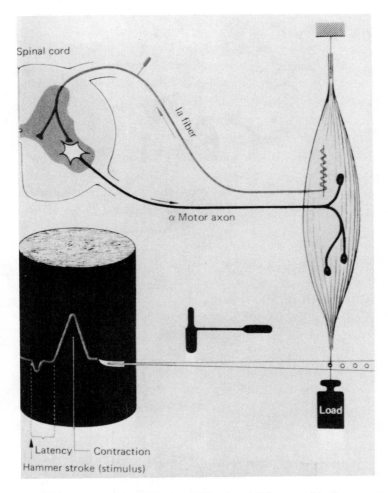

Figure 2. Reflex arc of the monosynaptic stretch reflex. A light step, with a hammer on the stylus recording muscle length (downward deflection of the trace on the recording paper) after a brief latency, produces contraction of the muscle. The reflex arc underlying this response is diagrammed, from the muscle spindles via the Ia fibers to the motorneurons and back to the muscle. (From Schmidt et al., 1978)

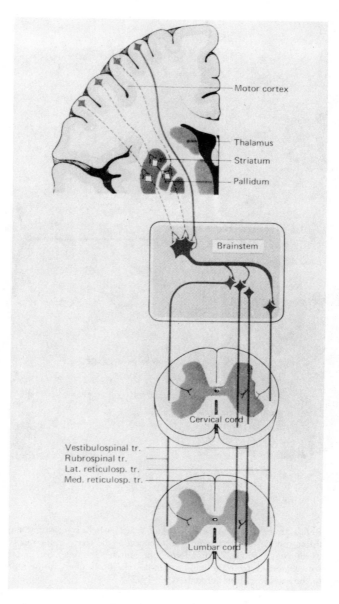

Figure 3. Schematic diagram of the courses of the most important extrapyramidal tracts from the supraspinal motor centers into the spinal cord. The neuron with a thick axon in the brainstem symbolizes the crossing of most of the extrapyramidal motor fibers to the opposite side at that level, and does not imply convergence. The pathways from motor cortex to basal nuclei are partly collaterals of the corticospinal tract and partly separate efferents. The details of connectivity among the brainstem structures involved in motor activity are extremely complicated; the representation here is greatly simplified. (From Schmidt et al., 1978)

4

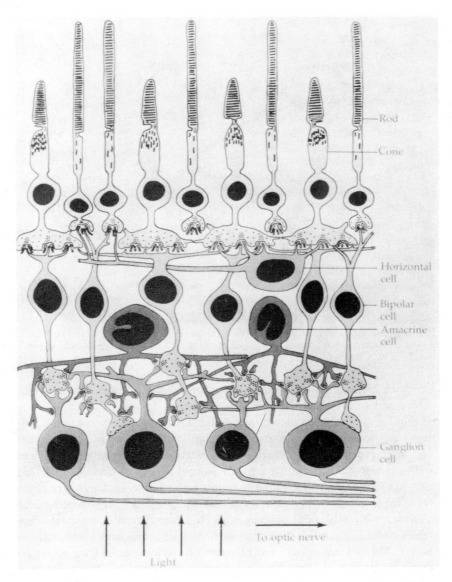

Figure 4. Organization of primate retina, after Dowling and Boycott (1966). (From Kuffler and Nicholls, 1977)

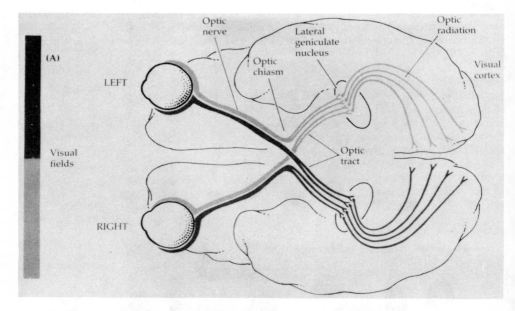

Figure 5. Outline of the visual pathways seen from below (base of the brain) in primates. The right side of each retina projects to the right lateral geniculate nucleus and the right visual cortex receives information exclusively from the left half of the visual field. (From Kuffler and Nicholls, 1977)

result in a sensation or an image, to afford choices, to generate abstractions, and to execute behaviors.

Similar to the very incomplete understanding of how we sense, integrate, and express information is our understanding of how we store it for later recall. When we seek manifestations of information storage within a nervous system, we logically would expect to find changes that persist. These changes may be manifest with biophysical measurements (i.e., involving the flux of ions across membranes), with biochemical assays, or by structural assessments. However they are manifest, they really must constitute a biological record of what is learned and remembered. These learning-induced changes are what remain long after the original stimulus patterns that produced them are gone. These changes, be they biophysical, biochemical, and/or structural are what give memory its physical reality.

How do we find them? And how do we reconstruct the process by which they occur? In the best of all possible worlds, we would trace the

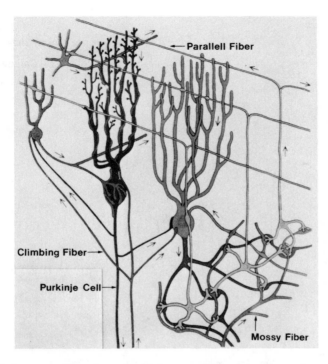

Figure 6. Interconnection of neurons in the cortex follows an elaborate but sterotypic pattern. Each Purkinje cell is associated with a single climbing fiber and forms many synaptic junctions with it. The climbing fiber also branches to the basket cells and Golgi cells. Mossy fibers come in contact with the terminal "claws" of granule-cell dendrites in a structure called a cerebellar glomerulus. The axons of the granule cells ascend to the molecular layer, where they bifurcate to form parallel fibers. Each parallel fiber comes in contact with many Purkinje cells, but usually it forms only one synapse with each cell. The stellate cells connect the parallel fibers with the dendrites of the Purkinje cell, the basket cells mainly with the Purkinje-cell soma. Most Golgi-cell dendrites form junctions with the parallel fibers but some join the mossy fibers; Golgi-cell axons terminate at the cerebellar glomeruli. Cells are identified in the key at lower left; arrows indicate direction of nerve conduction. (From Llinas, 1975)

information as it entered the nervous system (i.e., at the input stage); we would follow it as it coursed through all the integrative steps. Through neural pathways, we would watch the progressive transformations at critical sites within the pathways, and we would determine how these transformations altered information that exited the nervous system (i.e., at the output stage). The impracticality of even approximating such an analysis in a vertebrate nervous system, as suggested above, is over-

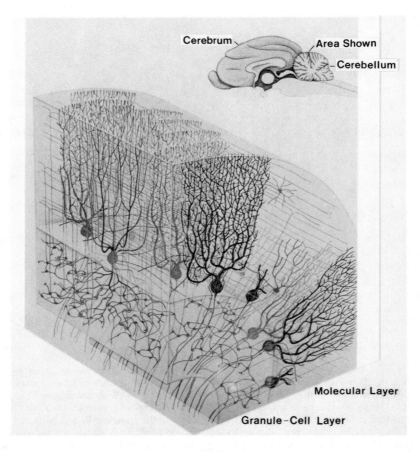

Figure 7. Architecture of the cortex of the cerebellum is diagrammed for a section of tissue from the brain of a cat. The location of the tissue section is indicated in the drawing at the top right; the same array of cells is repeated throughout the cortex. The cortex is organized around the Purkinje cells, whose somas, or cell bodies, define the border between the superficial molecular layer and the deeper granule-cell layer. In the molecular layer are the Purkinje-cell dendrites, which are arrayed in flattened networks like pressed leaves, and the parallel fibers, which pass through the dendrites perpendicularly. This layer also contains the stellate cells and the basket cells, which have similarly flattened arrays of dendrites. In the deeper layer are the granule cells, which give rise to the parallel fibers, and the Golgi cells, which are characterized by a cylindrical dendritic array. Input to the cortex is through the climbing fibers and mossy fibers; output is through the axons of Purkinje cells. (From Llinas, 1975)

whelming. Yet it is in the vertebrate nervous system, more particularly a mammalian nervous system, and to be precise the human nervous system, in which we wish to uncover the physical reality of memory.

Confronted with this dilemma, scientists over the last several decades developed several major experimental strategies. One strategy involved the use of lesions–ablations of critical brain areas to eliminate what might be a site essential for the storage of a particular memory. This approach was reinforced by a clinical strategy that depended on the pathologic examination of human brains to search for natural lesions that might account for memory deficits previously demonstrated in living patients. Clinicians for many generations made such observations, accumulating experience with a variety of pathologic lesions associated with a variety of clinical syndromes.

With the introduction of electrophysiologic techniques, new possibilities emerged. Microelectrodes could be placed in well-specified brain regions during surgery and used to inject very small amounts of current that might elicit memories. Patients in a conscious state and without pain were able to communicate remembrance of past experience in response to the microelectrode stimulation. More promising were the extracellular potentials that could be recorded and amplified by electrodes inserted into the brains of vertebrate species during the acquisition and retention of learning. Workers monitored changes of extracellularly recorded electrical activity (i.e., activity recorded outside of cells) as they were correlated with learning. Lesion studies were often coupled with recording measurements in efforts to localize memory storage sites.

Despite these efforts and these advances, because of the complexity of the vertebrate brain, mapping the involved neural circuits and their learning-induced modification has proved formidable and, to a considerable degree, elusive, although some very promising results have just recently become available (see Chapter 15).

Another research strategy involved biochemical intervention and measurements. This work at first centered largely on manipulations concerning protein synthesis. Animals given learning tasks were found to have differences in RNA metabolism, and inhibition of protein synthesis disrupted the retention of learned behavior. Later pharmacologic manipulations of neurotransmitter substances (which carry signals across synaptic junctions between neurons) and hormonal agents were shown to affect learning. Still elusive, however, were the critical mechanisms and local sites whereby such manipulations caused their effects. It was never clear, for instance, whether a change in protein synthesis was a cause or

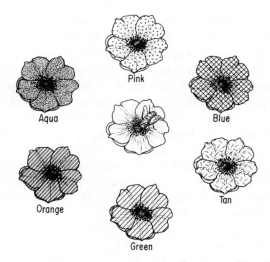

Figure 8. Bees learn to associate the color of a flower with food. (From Menzel and Erber, 1978)

an effect of learning, whether such a change actually stored information or simply passively reflected such storage elsewhere within the brain, and frequently, whether or not such changes also represented part of sensory, integrative, or motor physiology not at all unique to learning.

Still another strategy involved the use of learning in less evolved species as models for our own learning capacity. It was hoped that enough might be common to learning of simple and more complex animals that cellular insights in the former would have relevance for the latter. Among the earliest preparations, which are still useful today, were the locust and grasshopper, bees (Figure 8), and gastropod molluscs more familiar to us as snails. With these preparations, it was thought possible to combine electrophysiologic and biochemical analyses to arrive at complete descriptions of neural circuits and to localize the actual mechanisms for storage and recall of learned information that bore resemblance to, or shared features with, that which we know in our own conscious experience.

Bibliography

Alkon, D. L. (1980). Cellular analysis of a gastropod (*Hermissenda crassicornis*) model of associative leraning. *Biol. Bull.* 159:505–60.
Alkon, D. L., and Farley, J. (eds.) (1984). *Primary Neural Substrates of Learning*

and Behavioral Change. Cambridge University Press.

Bullock, T. H., and Quarton, G. C. (1966). Simple systems for the study of learning mechanisms. *Neurosci. Res. Prog. Bull.* 4:203–327.

Burrows, M., and Horridge, G. A. (1974). The organization of inputs to motoneurons of the locust metathoracic leg. *Philos. Trans. R. Soc. Lond. B. Biol.* 269:49–94.

Cowan, W. M. (1981). *Studies in Developmental Neurobiology: Essays in Honor of Viktor Hamburger.* Oxford University Press, New York.

Dowling, J. E., and Boycott, B. B. (1965). Neural connections of the retina: fine structure of the inner plexiform layer. *Cold Spring Harbor Symp. Quant. Biol.* 30:393–402.

(1966). Organization of the primate retina: electron microscopy. *Proc. R. Soc. London B* 166:80–111.

Flexner, J. B., and Flexner, L. B. (1969). Studies on memory: evidence for a widespread memory trace in the neocortex after the suppression of recent memory by puromycin. *Proc. Natl. Acad. Sci. U.S.A.* 62:729–32.

Hubel, D. H., and Wiesel, T. N. (1972). Laminar and columnar distribution of geniculo–cortical fibers in the macaque monkey. *J. Comp. Neurol.* 146:421–50.

Hyden, H., and Egyhazi, E. (1962). Nuclear RNA changes of nerve cells during a learning experiment in rats. *Proc. Natl. Acad. Sci. U.S.A.* 48:1366–73.

(1964). Changes in RNA content and base composition in cortical neurons of rats in a learning experiment involving transfer of handedness. *Proc. Natl. Acad. Sci. U.S.A.* 52:1030–5.

Kandel, E. R., and Schwartz, J. H. (1982). Molecular biology of learning: modulation of transmitter release. *Science* 218:433–43.

Kesner, R. P. (1982). Mnemonic function of the hippocampus: correspondence between animals and humans. In *Conditioning: Representation of Involved Neural Functions,* ed. by C. D. Woody, pp. 75–88. Plenum, New York.

Kuffler, S. W., and Nicholls, J. G. (1977). *From Neuron to Brain.* Sinauer, Sunderland, Mass.

Lasansky, A. (1971). Synaptic organization of cone cells in the turtle retina. *Philos. Trans. R. Soc. Lond. B Biol.* 252:265–81.

Llinas, R. R. (1975). The cortex of the cerebellum. *Sci. Am.* 232:56–71.

(1981). Electrophysiology of the cerebellar networks. In *Handbook of Physiology,* sect. 1, *The Nervous System,* vol. II, part 2, *Motor Control,* ed. by V. B. Brooks, pp. 831–76. American Physiological Society, Bethesda, Md.

Macagno, E. R., Lopresti, V., and Levinthal, C. (1973). Structure and development of neuronal connections in isogenic organisms: variations and similarities in the optic system of *Daphnia magna. Proc. Natl. Acad. Sci. U.S.A.* 70:57–61.

Menzel, R., and Erber, J. (1978). Learning and memory in bees. *Sci. Am.* 239:102–10.

Schmidt, R. F. (1978). *Fundamentals of Neurophysiology.* Springer-Verlag, New York.

Shepherd, G. (1979). *The Synaptic Organization of the Brain.* Oxford University Press, New York.

Stensaas, L. J., Stensaas, S. S., and Trujillo-Cenoz, O. (1969). Some morphological aspects of the visual system of *Hermissenda crassicornis* (Mollusca: Nudibranchia). *J. Ultrastruct. Res.* 27:510–32.

Thompson, R. F., Barchas, J. D., Clark, G. A., Donegan, N., Kettner, R. E., Lavond, D. G., Madden, J., IV, Mauk, M. E., and McCormick, D. A. (1984). Neuronal substrates of associative learning in the mammalian brain. In *Primary Neural Substrates of Learning and Behavioral Change*, ed. by D. L. Alkon and J. Farley, pp. 71–99. Cambridge University Press.

Thorpe, W. H. (1963). *Learning and Instinct in Animals*. Harvard University Press, Cambridge, Mass.

2
Simple systems

But does it learn? This is the crucial issue with which the investigator must first grapple in choosing an animal with a potentially "simple" nervous system, one with few enough cells to permit a point-by-point description (i.e., a wiring diagram). Implicit within this issue is the question: What is learning? We begin with what is learning as we know it. We learn the names of things, we learn places, we learn to recognize, we learn how to do things, we learn how to behave socially, we learn logical relationships. Sometimes when we recall what we have learned we express it by some action – a greeting, a stroke with a tennis racquet, or a sequence of finger and hand movements across a piano keyboard. Often our recall is an entirely internal process – we recall, but do nothing. So human learning and its recall need not involve behavior but only the *potential* for behavior. There is always a potential, for instance, to report to others what we have remembered.

Reflection on all possible examples of our memories leads to an obvious initial conclusion. Our memories are *always* complex. They never involve one bit of information in isolation of other bits. Rather, they consist of constellations of bits and the relationship of those bits in time and space. When we remember a face, we remember a nose, a mouth, forehead, etc., in relation to each other. The relationships of the parts generate the whole – the face is the total effect of the location in space of its various parts. Similarly, a forehand tennis stroke is really a sequence of movements, which linked together in exactly the right way in time, provide a total effect. The hand makes a certain grip, the shoulder rotates, the wrist is stiffened, etc. The stroke depends on the relation in time of many movements to each other as well as to the movement and the projected movement of the ball.

The universal complexity of our memories suggests the possibility of a unifying simplicity: What we remember is really the relationships between stimuli rather than the stimuli themselves and any given memory

13

is merely a particular collection or set of *relationships*. If we can understand how the nervous system stores and recalls relationships, maybe we can then gain insight into the process of learning and memory, in general. Much of the importance of conditioning originally defined by Pavlov derives from the fact that it provided a clear behavioral expression of an animal learning the temporal relationship between two stimuli: a bell and a piece of meat. Prior to the training experience the meat elicited salivation, whereas the bell did not. The bell was repeatedly presented shortly before the meat. Subsequently, the bell alone elicited salivation. As defined by the animal's new response to the sound of the bell, it had learned the relationship between the sound and the meat – the sound predicted the meat, and, thus, the animal, on hearing the bell, anticipated the meat. The defining features of this learned relationship – this association – might also define a basic unit of our own memory. Such units, multiplied a vast number of times, might together create a fabric of learned associations that together comprise a remembered experience.

One defining feature of Pavlovian or classical conditioning is its improvement with practice. The more times the meat and bell are presented together, the more likely the dog is to learn the relationship. Another feature is the specificity of the learned relationship for only the stimuli presented. Another stimulus, a light flash, for example, which is presented with some delay after the temporally related bell and meat, does not become associated with the bell or meat. Still another feature concerns the persistence of the learned association – it lasts for many weeks or even years. These and other hallmarks of classical conditioning can also be observed for other forms of conditioning such as operant conditioning, whereby an animal learns to behave in a manner to achieve some end. These hallmarks have, in fact, been identified for a myriad of examples of learning, all of which concern relationships in time between stimuli or aggregates of stimuli. Obviously, in addition, there may be properties that uniquely characterize a particular form of associative learning. But those features, which transcend the particularities of the form, point to the possibility of general underlying behavioral principles and causal cellular mechanisms as well.

Within the context that learning must involve the relationships of stimuli, we can begin to sort out behavioral and cellular phenomena that may or may not provide useful subjects of learning studies. A whole class of phenomena that do not concern the relationships of stimuli can be described as nonassociative. Repeated or prolonged presentation of only one stimulus results in some change of subsequent stimulus responses. Sensory adaptation is one example of such a nonassociative phenomenon. Bright-light stimulation of photoreceptors will cause "light adaptation." For many minutes, even hours, after substantial exposure to bright light the photoreceptors will be less responsive to subsequent presentations of a light stimulus – the photoreceptors will generate a smaller electrical signal in response to a test light. The dimmest light detectable by the light-adapted photoreceptor is also larger than that detectable by the dark-adapted cell. Our pressure receptors also adapt so that, although we are at first aware of our body contact with a piece of furniture, we soon lose consciousness of this contact. Similar to sensory adaptation, and perhaps even a type of sensory adaptation, is habituation. Touching an animal's limb may elicit a withdrawal response. Repeated touching can be associated with progressively diminished limb withdrawal and perhaps even none at all. The limb withdrawal response to touch has habituated. Habituation is characteristically reversible. It may reverse spontaneously in a manner similar to the way sensory cells regain their sensitivity after adaptation. Habituation also may reverse when the animal is generally aroused or sensitized by the presentation of another strong second stimulus. The general facilitatory effect of sensitizing stimulation on responsiveness to other stimuli is also a nonassociative behavioral phenomenon and is itself reversible. We are all familiar with the sensitizing or arousing effects of a shower on rising. It makes us more alert, more aware in general of what is happening around us.

Nonassociative behavioral changes have been extensively analyzed for a gastropod mollusc, a sea snail, called *Aplysia californica*. Repetitive touching of the *Aplysia* appendage, called the mantle, results in a decrease in the amplitude of and probability for a reflex withdrawal of another appendage called the gill. This reduction of the gill withdrawal response due to repeated presentations of a single stimulus (touch) is called habituation. It can last for weeks if the animal's sensory input is carefully controlled during maintenance conditions. If, however, the animal is exposed to a single presentation of a noxious stimulus such as "pinching" the "neck" region, the habituation is abolished. Application of electric shock to the "tail" region of the animals also abolishes the habituation. Presentations of either of these two stimuli, "pinching" or "tail" shock, will, in a nonhabituated animal, increase responsiveness of gill withdrawal as well as increase responsiveness to stimuli in general. This generalized increase of responsivity is an example of "sensitization" or arousal.

Because nonassociative phenomena, such as sensory adaptation, habituation, and sensitization, are so fundamentally different from learning as we know it, it is important to distinguish their presence and influence within behavioral models being used to investigate biological mechanisms of learning and memory. This is not to say that some physiology could not be shared by associative learning and nonassociative behavioral effects. This is not to say that sensory adaptation, habituation, and sensitization could not participate or affect the process whereby we encode a true association. It is to say, however, we cannot in any way assume, or perhaps even expect, basic similarities in the way the nervous system provides for such very different forms of behavioral modification. In fact, as will become apparent from what is known about the encoding of an association by a neural system, the cellular mechanisms are indeed quite distinct from those mechanisms delineated thus far for sensory adaptation, habituation, and sensitization.

The search for preparations that learn and are amenable to a cellular analysis has uncovered many interesting possibilities. The bee, for example, provides one of the most elegant and well-studied examples of associative learning (i.e., learning of stimulus relationships by an invertebrate). The bee learns to associate a color with the olfactory stimulation of nectar. Depending on the bee's location in a natural setting, it will encounter with a high frequency flowers of various colors and color combinations. With sufficient experience with particular flowers, the bee learns the relationship in time of seeing the color cues and, subsequently, encountering the flower's olfactory signals and its nectar (Figures 8 and 9). Having learned this relationship, the bee will guide its movement in the direction of those colors even in the absence of the flowers. For the bee the color has taken on the meaning of the olfactory signals arising from the flower. Prior to the training experience, an olfactory signal from a flower elicited stereotypically guided movement – an inborn, reliable response of a signal with unequivocal meaning for the bee. Also prior to the training experience, the color did not have unequivocal meaning and did not elicit a reliable response – rather the color had a more neutral value. Once the relationship of the color and the olfactory signal was learned, the color (as defined by the bee's behavior) assumed the value of the olfactory signal – the color was now defined (for the bee) by its relationship to the flower's olfactory cue.

Exactly the same process occurs with Pavlov's dog. The smell of meat has unequivocal value for the dog and reliably elicits salivation. Prior to training, the sound of the bell is a relatively neutral stimulus. Having

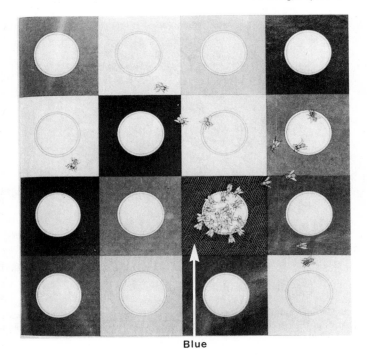

Blue

Figure 9. Color test made by Karl von Frisch in 1910 established the ability of bees to associate colors with food and to remember a color thus learned. He put a dish containing a honey solution on a table. Bees were attracted to it. Then he put the dish on a piece of blue cardboard, so that blue was the color the bees saw as they approached the dish and sucked the honey. Next, von Frisch surrounded the blue cardboard with other cards of the same size but colored white, black, or shades of gray. Each card, including the blue one, bore an empty dish. The bees had been conditioned to expect food at the table, and so they continued to come. Most of them looked only at the blue card. (From Menzel and Erber, 1978)

learned the relationship between the sound of the bell and the smell of the meat, the bell takes on (as behaviorally defined) the value of the meat and itself reliably elicits salivation.

It is remarkable that nature has preserved the essential form of a learned relationship between stimuli in spite of the dramatic differences that arose through the evolution of animals as diverse as the bee and the dog. It is perhaps equally remarkable that the same conservation of the learning process is manifest even in the associative learning of a snail. As might be expected, nonassociative behavioral influences such as sensory adaptation, habituation, and sensitization are readily demonstrated for

Figure 10. The nudibranch mollusc *Hermissenda crassicornis*. Note the small block spot (the right eye) at the base of the lower rhinophore. (From Alkon, 1980 – picture provided by L. M. Golder)

such animals. It was essential, therefore, in demonstrating associative learning for the snail to rule out the possibility that the learned stimulus relationship was not some combination of nonassociative behavioral changes – rather than a true association. This has now become possible for the snail called *Hermissenda crassicornis* (a nudibranch mollusc) (Figure 10).

Rotation of *Hermissenda* or shaking (which more closely resembles oceanic turbulence) causes a reliable and robust response – the animal will contract its muscular undersurface also known as its "foot" (Figure 11). This contraction effects a "clinging" (Figure 12) that enables the animal to adhere to surfaces on which it moves. The value of rotation for *Hermissenda* is unequivocally negative – the animal responds stereotypically to reduce the magnitude and effect of the stimulus. By contracting its foot and thereby clinging to a surface, *Hermissenda* will experience less disruptive movement and will minimize the probability of damage to its external appendages.

A gradient of light (as created by a light source placed at some distance from the animal, for instance) has a less stereotyped effect on

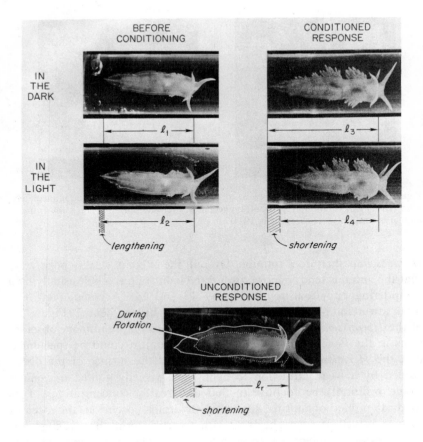

Figure 11. Photographic representation of classically conditioned responses of the *Hermissenda* foot. Bottom panel, an overlay of two photographs taken in the dark; solid white line, the outline of the foot 1 s before rotation; dashed line, the outline of the foot after 3 s of rotation at 97 rpm. Upper panels, comparisons of lengths in the light to lengths in the dark before and after conditioning with paired light and rotation stimuli. l_1, length in the dark before training; l_2, length in the light before training ($l_2 > l_1$ = lengthening); l_3, length in the dark during retention of learned behavior; l_4, length in the light during retention ($l_3 > l_4$ = shortening); l_r, length during rotation was always smaller than before rotation began. (From Lederhendler et al., 1986)

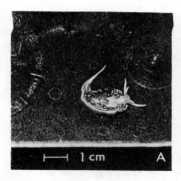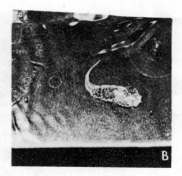

Figure 12. Response of *Hermissenda* to rotation immediately before (A) and immediately after (B) 20 s of rotation at 200 rpm. Immediately after and during rotation the entire animal is contracted. (From Alkon, 1974)

Hermissenda than does rotation. Overall, the effect is weakly positive in that the animals tend to move toward the light source, albeit usually by a meandering circuitous route (Figures 13 and 14). Accompanying this weak attraction of light for the animal is a slight lengthening (Figure 12) of the *Hermissenda* undersurface or foot. Based on preliminary observations in the Pacific (off the coast of California), it could be speculated that the *Hermissenda* naturally move toward the surface of the ocean guided by a weak attraction to light. Such visually guided movement would ordinarily be adaptive in that the greatest concentrations of its food, the microorganisms known as hydroids, occur at the ocean's surface as well. If, as might occur with a severe storm, there is considerable oceanic turbulence, it would begin to stimulate the *Hermissenda* as they moved toward the shallower depths in response to the prior onset of stimulation by a light gradient. Thus, light onset would precede turbulence onset as might be approximated by pairing light and rotation in the laboratory. The animal might repeatedly encounter the turbulence preceded by the onset of light and eventually learn not to go to the light and thus the ocean surface. This would have adaptive value in that the animal can survive for weeks with little or no food but cannot survive damage resulting from prolonged buffetting at the ocean's surface.

During a typical training regimen in the laboratory, the onset of a light step is repeatedly followed by the onset of rotation. This pairing of light and rotation is repeated at 2-min intervals for approximately $1\frac{1}{2}$ h on 3 successive days. On days and weeks following such training, the light is again presented alone (i.e., without rotation). The light now elicits a new

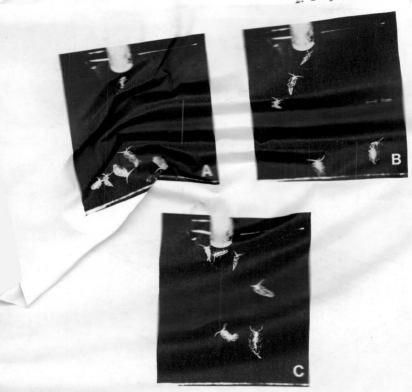

Figure 13. Response of animals in an experimental chamber to the test light spot. (A)–(C) Animals move toward and into the light spot within 20 min. The spot (indicated by dashed lines) is obscured by a stroboscopic flash in (B) and (C). Calibration: 1 cm. (From Alkon, 1974)

behavioral response – one that closely resembles that of rotation. After repeated pairings of light and rotation (Figure 15), light, as rotation before training, causes the *Hermissenda* foot to contract rather than to lengthen (Figures 12 and 16). Light, as defined by the animal's behavioral response, has taken on the meaning of rotation. Accompanying foot contraction in response to light is reduced velocity of movement toward the light source and thus reduction of the weak positive effect of a light gradient (Figure 15). The behavior of *Hermissenda* after training strongly suggests that a relationship between light and rotation has been learned. What other evidence demonstrates that, in fact, it was the relationship that was learned, that an association has been formed? Such evidence was provided by a series of control experiments that not only prove that

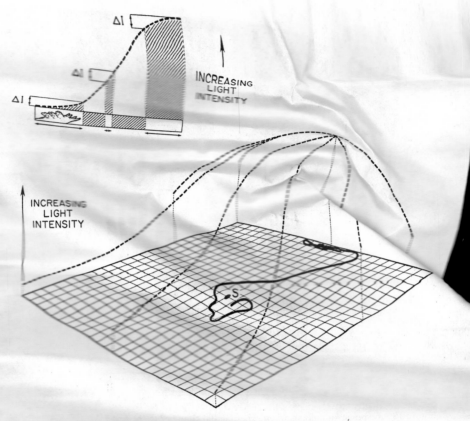

Figure 14. Tracing of a single mucus trail left behind by a positively phototactic *Hermissenda crassicornis*. The trail is superimposed in a 30° projection onto an open field aquarium illuminated from one side (see Lederhendler et al., 1980). This individual started in the center of the field (S) and move toward the more illuminated end. The dashed lines represent the intensities of light reaching the animal at different points in the field. *Inset*: A profile of the illumination gradient showing the distance an animal would have to move to experience a constant increment in light intensity (ΔI) changes for different parts of the gradient. This profile is similar to the stimulation experienced by *Hermissenda* tested in tubes. Scale: Each square in the open field is 1 cm². The inset is not to scale. (From Lederhendler and Alkon, 1986)

a true association between light and rotation is learned by these snails, but also reveal the defining features of associative learning as observed for vertebrate species. The conditioned response, the learned light-elicited foot contraction as well as the accompanying reduction of phototactic movement, was produced by training and optimal learning occurred when light preceded rotation by a limited temporal interval (0.5–1.0 s)

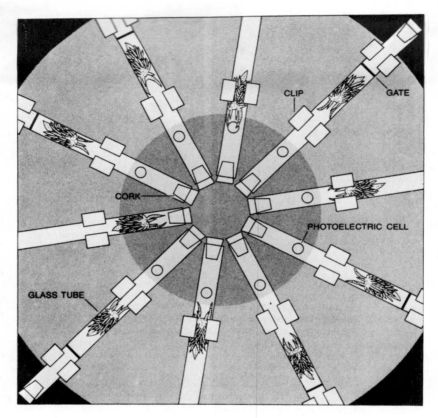

Figure 15A. Snails are trained and tested in glass tubes on a turntable. First, each snail's phototactic response to light before training is measured: The time the snail takes to reach a spot of light (darkly shaded center area) is recorded automatically when the animal reaches a photoelectric cell. Then the snails are trained by being rotated while confined to the outer end of the tube and thus subjected to a centrifugal force that is sensed by the statocysts. For one group the rotation is precisely paired with a 30-s period of light; various control groups either are not trained or are subjected to light or rotation alone or to the two stimuli alternately or at random. Finally, the snail's velocity of movement toward light is timed again to determine the effect of training. (From Alkon, 1983)

with paired light and rotation. The conditioned response was not produced by training with light and rotation stimuli, which occurred with no fixed temporal relationship with each other (Figure 15), when rotation preceded light, or when rotation followed light with too long a delay. Nor did a learned reduction of phototactic movement follow light and rotation that alternated during training. Similarly, training with repeated presentation of either the light or rotation alone during training did not

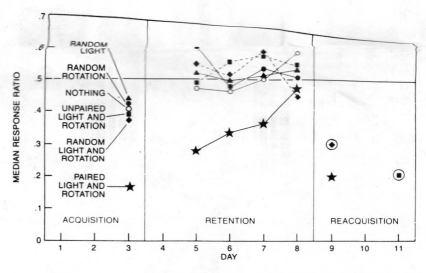

Figure 15B. Conditioning is demonstrated by these data. The "response ratio" measures the suppression of the phototactic response. A ratio of 0.5 means, in effect, that there has been no change in the velocity of movement toward light; a lower ratio signifies a reduction in the velocity. Animals subjected to paired light and rotation (stars) are conditioned after 3 days of training; they move more slowly toward light. They "forget" that behavior slowly but reacquire it after 1 day of retraining. Control animals show no significant change in behavior (black). Yet they too can be conditioned (solid circles) when subjected to paired stimuli. (From Alkon, 1983)

result in acquisition of the learned response nor, in fact, any other prolonged behavioral change. Training with rotation preceding the light or with rotation following light onset by too great a delay also did not cause learning. No prolonged sensory adaptation, habituation, or sensitization occurred after training with the light and rotation in isolation of each other. Thus, the learned response following light and rotation pairings cannot be attributed to some particular combination of these nonassociative behavioral changes. Furthermore, addition of unpaired presentations of either the light or the rotation during training with repeated pairs of light and rotation degrades the acquisition of the learned response. If the learned response were due to sensory adaptation, habituation, or sensitization, additional unpaired presentations would enhance, not hinder, the learning.

Other features characterized both the *Hermissenda* associative learning and that of vertebrates. *Hermissenda* associative learning increased as a function of practice (i.e., it showed acquisition). Its retention (ordinarily

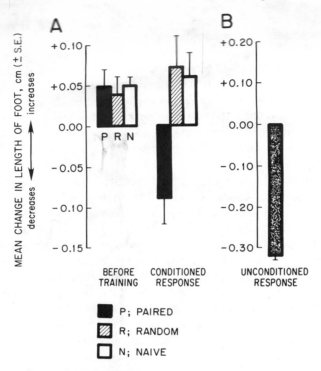

Figure 16. Classical conditioning of a new response to light in *Hermissenda*. (A) Before training, foot length increased in response to light onset. Random (▨) and naive (□) groups continue to show light-elicited lengthening, but paired (■) animals became conditioned to shorten the length of the foot in response to a light stimulus. (B) Unconditioned response after 6 s of rotation. The CR is about 28% of the UCR. Note the difference in scale between (A) and (B). (From Lederhendler et al., 1986)

for weeks after training) was diminished by repeated presentation (after training) of light stimuli alone (i.e., it showed extinction). At the end of retention, when the animal had "forgotten" the learned response, it relearned the conditioned response more quickly, with fewer paired stimulus presentations – it showed what has been termed "savings" for vertebrate learning. Finally, the response to light, but not to other stimuli that were not presented during training, changed with the learning, that is, it was "stimulus-specific."

Parallels between *Hermissenda* learning of an association and that of vertebrates were apparent in still another context. *Hermissenda* collected from the Pacific vary considerably in their movement toward a light

source. Also variable is the ease with which such "wild" *Hermissenda*
learn to associate light with rotation. In fact, the collected nudibranchs
tend to fall into two classes: one that shows fast, the other slow, learning
ability. Aside from innate physiologic differences that could contribute
to differences in learning ability, we might expect wide variation in the
history of environmental stimulation encountered by the animals in their
natural habitat. When the animals arrive at the laboratory from the
Pacific, they may have already experienced some association of light and
oceanic turbulence (and thereby had their light-induced movements
modified).

We predicted that variation in the animal's phototactic and learning
behavior would be reduced by raising laboratory strains and thus controlling both its genetic and environmental history. In fact, this proved to
be true for laboratory-reared *Hermissenda* (i.e., animals derived from
eggs fertilized in the laboratory). For these animals, measurements of the
response to light as well as the acquisition of the learned association
were much more uniform than for animals from the sea.

But is it learning? Certainly the learning of the snail *Hermissenda* is
not identical to the learning of a rabbit, a dog, or ourselves. It would be
unreasonable to expect that a nervous system with vastly less capability
for sensing, integrating, and implementing information is going to associate stimuli with the same temporal resolution, the same potential to
generalize to other related stimuli, or the same potential to build abstractions from discrete learned associations. As will be discussed later, not
only is the number of neurons in a snail's nervous system many orders of
magnitude smaller than those of an animal such as a rabbit, the speed
with which signals move along the long processes of neurons is slower
for the snail; the delay with which the signal releases a chemical
messenger at communicating junctions with other neurons is greater. The
location and distribution of these junctions is dramatically different for
the snail and the rabbit. So, there are essential differences in the design
of the neural machinery at these different levels of evolution. And, these
essential differences must have some expression in behavioral phenomena including those involved in learning. Yet there are many important
aspects of the neural machinery that have not changed in the course of
extensive evolution. There is much that is common to the nervous
systems of snails, rabbits, dogs, and humans. Enough is in common, as
was shown for *Hermissenda*, bees, dogs, and humans, to also have
behavioral expression. No one of the characteristics identified for
Hermissenda associative learning is sufficient to indicate that such learn-

ing may serve as a convincing model for vertebrate classical conditioning. It is the *list* of characteristics, which really includes all of the critically involved features of vertebrate classical conditioning, that ultimately offers us conviction of the usefulness of the snail model of associative learning. It cannot be overemphasized that simply finding a few features (such as a dependence on pairing and/or prolonged retention) does not mean that another example of snail associative learning has been found. Does the behavior show extinction, does it show "savings," does it show stimulus specificity, does it show uniqueness from, and independence of, nonassociative phenomena, does it show a true transference of meaning from the unconditioned stimulus to the conditioned stimulus? These and other questions must be asked before an associative model has indeed been established, and before the type of associative learning can be identified.

With an established model, however, there is an exciting potential. The potential exists that if enough is held in common behaviorally and neurologically, then the way the information is stored – the cellular language used for encoding and representing the relationship between stimuli – may also be common to the snail and higher vertebrates. Indeed this potential, this promise, has, at least in one clear instance, been confirmed and will be described in subsequent chapters.

Bibliography

Alkon, D. L. (1974). Associative training of *Hermissenda*. *J. Gen. Physiol.* 64:70–84.
   (1980). Cellular analysis of a gastropod (*Hermissenda crassicornis*) model of associative learning. *Biol. Bull.* 159:505–60.
   (1983). Learning in a marine snail. *Sci. Am.* 249:70–84.
Crow, T. J., and Alkon, D. L. (1978). Retention of an associative behavioral change in *Hermissenda*. *Science* 201:1239–41.
Erber, J. (1984). Response changes of single neurons during learning in the honeybee. In *Primary Neural Substrates of Learning and Behavioral Change*, ed. by D. L. Alkon and J. Farley, pp. 275–84. Cambridge University Press.
Gormezano, I. (1984). The study of associative learning with CS–CR paradigms. In *Primary Neural Substrates of Learning and Behavioral Change*, ed. by D. L. Alkon and J. Farley, pp. 5–24. Cambridge University Press.
Kandel, E. R. (1976). Cellular Basis of Behavior: An Introduction to Behavioral Neurobiology. San Francisco, Freeman & Co.
Lederhendler, I., and Alkon, D. L. (1986). Reduced withdrawal from shadows by associative conditioning of the mollusc *Hermissenda*: a direct behavioral analogue of photoreceptor responses to brief light steps. *Behav. Neural Biol.* In press.

Lederhendler, I., Gart, S., and Alkon, D. L. (1986). Classical conditioning of *Hermissenda*: origin of a new response. *J. Neurosci.* 6:1325–31.

Lindauer, M. (1970). Lernen und Gedachtnis – Versuche an der Honigbiene. *Naturwissenschaften* 57:463–7.

Menzel, R. (1984). Short-term memory in bees. In *Primary Neural Substrates of Learning and Behavioral Change*, ed. by D. L. Alkon and J. Farley, pp. 259–74. Cambridge University Press.

Menzel, R., and Erber, J. (1978). Learning and memory in bees. *Sci. Am.* 239:102–10.

Pavlov, I. P. S. P. (1927). *Conditioned Reflexes* (translated by G. V. Anrep). Oxford University Press, London.

Rescorla, R. (1984). Comments on three Pavlovian paradigms. In *Primary Neural Substrates of Learning and Behavioral Change*, ed. by D. L. Alkon and J. Farley, pp. 25–45. Cambridge University Press.

Thompson, R. F., and Spencer, W. A. (1966). Habituation: a model phenomenon for the study of neuronal substrates of behavior. *Psychol. Rev.* 73:16–43.

von Frisch, K. (1967). *The Dance Language and Orientation of Bees*. Cambridge University Press.

3
Defining neural systems

Given an invertebrate model of vertebrate classical conditioning, how do we exploit it to uncover cellular mechanisms? What is the game plan? Using the snail *Hermissenda* as an example, let us review the steps taken to arrive at cellular and subcellular mechanisms by which one stimulus becomes associated with another – by which a temporal relationship between stimuli is learned. We begin by mapping out the nervous system – by putting together what is in essence a wiring diagram consisting of neural elements, which are wired together and which are connected. By recording electrical signals from these elements, we can monitor the flow of these signals from one cell to the next.

Neural electrical signals are recorded by amplifying very small potential differences measured by microelectrodes inserted into the cells. A microelectrode consists of a conducting material such as a silver wire placed within an electrolyte solution, which in turn fills a capillary tube (Figure 17). The capillary tube is shaped (by heating) into a very fine tip of much smaller dimension ($< 0.5 \times 10^{-6}$ m) than the diameters of the cells themselves ($2–200 \times 10^{-6}$ m). Microelectrodes inserted into several neurons simultaneously monitor the timing of a signal in one cell in relation to the signal received by a connected cell. A large voltage change (an action potential) travels along a neural fiber without decrement to a site of contact between cells (called a synaptic junction) and causes the release of a chemical messenger (a neurotransmitter) that diffuses across the contact site to the process of a second cell (Figure 18). The neurotransmitter, upon reaching the surface of the second cell, interacts with specialized molecular complexes known as receptors. This messenger–receptor interaction in turn causes the generation of an electrical signal that now moves along the structure of the receiving cell and in turn can trigger other action potentials (Figure 19). This process of one cell's action potential sending a chemical message, which in turn influences another cell's action potential, defines a synaptic relationship.

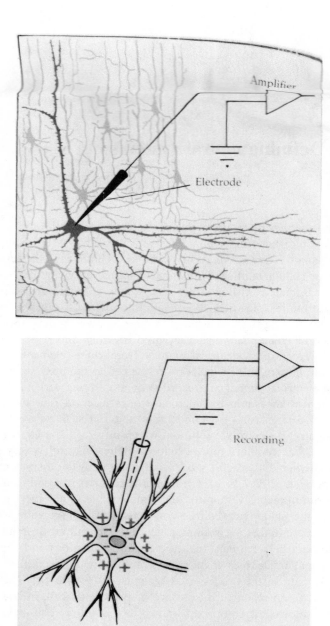

Figure 17. (A) Extracellular recording with a fine wire electrode. The electrode tip has been drawn close to a nerve cell in the cortex. (B) Intracellular recording. The tip of a microelectrode has been inserted into a nerve cell. In a neuron that is at rest, there is a potential difference of about 70 mV; the inside is negative with respect to the outside. (From Kuffler and Nicholls, *From Neuron to Brain*, 1977)

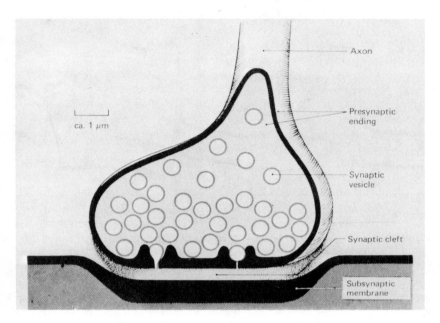

Figure 18. Diagrammatic section through a chemical synapse. All the structural elements important in synaptic transmission are shown. The diameter of the synaptic vesicles and the width of the synaptic cleft are greatly exaggerated with respect to the other elements of the synapse. (From Schmidt et al., 1978)

Mapping a neural system requires establishing which synaptic relationships are present in every adult member of a species (i.e., determining the genetically programmed synaptic contacts between cells). To accomplish this in a snail's nervous system, it is first necessary to recognize the same cells in one animal after another and then, using microelectrodes, to measure the reproducible features of the electrical interactions between the cells. Once a synaptic relation for two identified cells is established with a high degree of statistical confidence (i.e., with an adequate sample size and with sufficient reproducibility), another major issue must be confronted. Is the observed interaction between the two cells direct or are there other intervening neural elements through which the signals are transmitted between the two cells? There are many functional criteria that help resolve this issue, such as the delay between the action potential of one cell and the signal received by the other cell. In the end, however, the only conclusive demonstration of a direct contact between cells is provided by visualizing the structural nature of

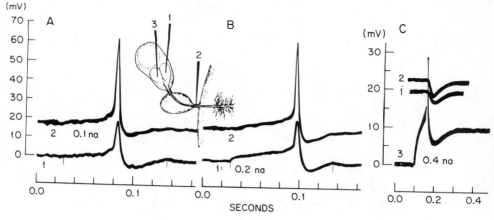

Figure 19. An action potential recorded from one cell produces a synaptic potential recorded from a second cell. Two microelectrodes were inserted, respectively, in the soma and axon of one receptor (microelectrodes 1 and 2) and a third microelectrode was inserted in the soma of a second receptor. (A) Responses recorded from the soma and axon of the same cell. The distance between the two electrodes was 80–100 μm. Responses to a step of depolarizing current through an axonal electrode are also shown. (B) Response to a step of current through the soma electrode. In either case the spike is larger and appears earlier in the axon. (C) A step of depolarizing current through this third electrode evokes a spike in one receptor and a hyperpolarizing synaptic potential in the other. The synaptic potential is larger and has shorter time to peak in the axon. (From Alkon and Fuortes, 1972)

this contact between cells with the great magnifying power of the electron microscope.

Intracellular recording from synaptically related neurons and electron microscopic techniques have been combined to arrive at a comprehensive description of the *Hermissenda* neural organization or wiring diagram within the neural pathway that is responsive to the unconditioned stimulus, rotation, and within the pathway responsive to the conditioned stimulus, light. It was also necessary to thoroughly understand the wiring diagram for the interaction or convergence of these two pathways. Before considering the details of these wiring descriptions, let us continue with our overview of experimental strategy.

With a thorough knowledge of the relevant wiring diagrams, we then examine with our microelectrodes what changes within the wiring of only the conditioned animals. These electrophysiologic changes are correlated with behavioral changes that occur during the acquisition and retention of the learned association between light and rotation. It then becomes necessary to find out which of the changes are primary and which are

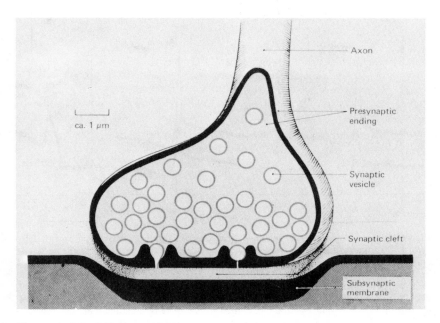

ca. 1 μm

Axon

Presynaptic
ending

Synaptic
vesicle

Synaptic cleft

Subsynaptic
membrane

Figure 18. Diagrammatic section through a chemical synapse. All the structural elements important in synaptic transmission are shown. The diameter of the synaptic vesicles and the width of the synaptic cleft are greatly exaggerated with respect to the other elements of the synapse. (From Schmidt et al., 1978)

Mapping a neural system requires establishing which synaptic relationships are present in every adult member of a species (i.e., determining the genetically programmed synaptic contacts between cells). To accomplish this in a snail's nervous system, it is first necessary to recognize the same cells in one animal after another and then, using microelectrodes, to measure the reproducible features of the electrical interactions between the cells. Once a synaptic relation for two identified cells is established with a high degree of statistical confidence (i.e., with an adequate sample size and with sufficient reproducibility), another major issue must be confronted. Is the observed interaction between the two cells direct or are there other intervening neural elements through which the signals are transmitted between the two cells? There are many functional criteria that help resolve this issue, such as the delay between the action potential of one cell and the signal received by the other cell. In the end, however, the only conclusive demonstration of a direct contact between cells is provided by visualizing the structural nature of

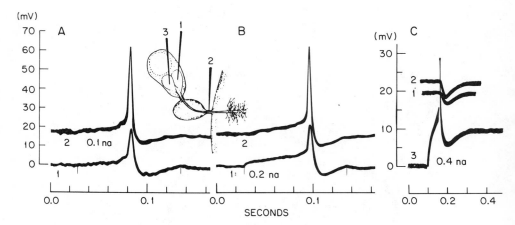

Figure 19. An action potential recorded from one cell produces a synaptic potential recorded from a second cell. Two microelectrodes were inserted, respectively, in the soma and axon of one receptor (microelectrodes 1 and 2) and a third microelectrode was inserted in the soma of a second receptor. (A) Responses recorded from the soma and axon of the same cell. The distance between the two electrodes was 80–100 μm. Responses to a step of depolarizing current through an axonal electrode are also shown. (B) Response to a step of current through the soma electrode. In either case the spike is larger and appears earlier in the axon. (C) A step of depolarizing current through this third electrode evokes a spike in one receptor and a hyperpolarizing synaptic potential in the other. The synaptic potential is larger and has shorter time to peak in the axon. (From Alkon and Fuortes, 1972)

this contact between cells with the great magnifying power of the electron microscope.

Intracellular recording from synaptically related neurons and electron microscopic techniques have been combined to arrive at a comprehensive description of the *Hermissenda* neural organization or wiring diagram within the neural pathway that is responsive to the unconditioned stimulus, rotation, and within the pathway responsive to the conditioned stimulus, light. It was also necessary to thoroughly understand the wiring diagram for the interaction or convergence of these two pathways. Before considering the details of these wiring descriptions, let us continue with our overview of experimental strategy.

With a thorough knowledge of the relevant wiring diagrams, we then examine with our microelectrodes what changes within the wiring of only the conditioned animals. These electrophysiologic changes are correlated with behavioral changes that occur during the acquisition and retention of the learned association between light and rotation. It then becomes necessary to find out which of the changes are primary and which are

4

Neural systems of Hermissenda

Neural systems

The details of the wiring diagram within and between the *Hermissenda* visual and statocyst pathways (see Appendix I for a more complete discussion) will not be extensively described in the brief summary that follows. As they become relevant to our later discussion, however, such details will be considered. A few general statements can be made about our knowledge of these pathways. First, because of the remarkably small number of cells involved, the functional organization of the cells that sense and integrate visual and vestibular stimuli in the environment is probably known with a degree of completeness, which is not frequently possible. Comparably defined sensory systems include the peripheral visual system of the horseshoe crab *Limulus* and the pressure–touch sensory system of the leech. The pressure-sensitive sensory cells' interactions with motorneurons of *Aplysia*, another snail, have also been throughly analyzed, although less is known about the interactions between *Aplysia* sensory cells and between sensory cells and interneurons.

The organization of visual–vestibular neural connections in *Hermissenda* provides an instance of intersensory wiring (i.e., the neural organization responsible for convergence of signals from two distinct sensory modalities) being comprehensively understood. Again, the number of neurons, their accessibility to electrophysiologic and histologic examination, and their recognizability have made this understanding possible. In fact, it was this well-defined intersensory convergence that motivated the subsequent experiments designed to train the animal to learn an association between stimuli mediated by these converging pathways. It was hypothesized that convergence of pathways, responsive to natural stimuli occurring within the animal's environment, must precede the potential, and the subsequent realization of that potential to learn an association. This hypothesis was derived from reflections on human learning.

Humans can associate at will any bit of information that can be sensed or perceived with any other sensed bit – even if these bits have no perceived relationship such as two arbitrarily chosen numbers. Humans can make such associations within seconds or less (too brief a time to grow new neural pathways, that is, growth involving significant movement of neuronal branches – new connections between neurons that allow convergence of electrical signals representing stimuli to be associated). The implication of these considerations of human associative learning is that the convergence necessary to associate any perceived stimulus with any other perceived stimulus already exists within the fully developed human brain. Thus, when convergence can be found in a lower organism, it may mediate and define the potential for associative learning.

The experimental strategy of associatively training an animal with stimuli, to which the animal's neuronal pathways are responsive in its natural environment and which converge at demonstrated locations within these pathways, is in contrast to another approach frequently taken with invertebrates. With this other approach, stimuli such as electric shocks, which are not ordinarily encountered and which do not flow along discrete pathways, are used for training. Electric shock applied to a snail's tail or to its head does not just immediately affect only a few particular sensory neurons. The shock spreads diffusely, traveling through the skin to nonneural, as well as neural, tissues. Therefore, it is almost impossible to trace step-by-step the particular neurons (and neurons to which they are synaptically connected) that respond to and transmit the information about the shock and how this transmitted information converges with signals generated by another stimulus with which the shock is to be associated. Furthermore, the following questions must be asked: What possibility is there for preservation during the course of evolution of a mechanism that mediates a behavioral and neural change produced by a stimulus never encountered by the animal in its natural habitat? What relevance to natural associative learning does "learning" that involves such a shock have? The answers to these questions will only become available after many years of research, when mechanisms of associative learning have been unequivocally identified for both vertebrates and invertebrates. Therefore, to maximize the likelihood of uncovering learning mechanisms that *are* physiologic, *are* conserved during evolution, and *are* common to many species, it is more reasonable to employ stimuli that resemble those naturally encountered, are processed by cells designed to transduce and transmit information about them, and meet at specific sites within the nervous system (the specific loci of convergence).

Another general statement about the visual and vestibular neural organization in *Hermissenda* concerns the paths that electrical signals (initiated by visual and vestibular stimuli) take as they travel through the nervous system from the input stage to the output stage – from sensory cells to motorneurons that control movement. It is already clear that neither the visual nor the vestibular signals follow along single chains of neurons linked together by synaptic junctions. After the information is first processed in each of the two sensory organs, the eyes and the statocysts (primitive vestibular organs), it is *distributed* to several different neuronal chains. This distribution of visual and vestibular information allows it to be used for different functions. Light and patterns of light, as they affect distinct neuronal chains, can cause generalized arousal of the animal, can cause orientation and guided movement, can cause lengthening of the animal's foot, and can cause brisk avoidance of moving edges of shadows. Some of these functions may occur in a coordinated fashion; others occur in isolation of each other.

The *Hermissenda* eye is constructed in such a way that light can stimulate the light-sensitive portion of each of the five photoreceptors separately. The light-sensitive portions, known as rhabdomes, of each of the five photoreceptors (in each of the two eyes) have particular locations around a single spherical lens (Figures 20 and 21). Darkly pigmented cells (unresponsive to light and without electrical or synaptic signals) envelope each photoreceptor and rhabdome in such a way as to prevent light from stimulating the photoreceptor except as it (light) comes through the lens (Figures 21 and 22). This structural arrangement of the eye helps it detect differences in light intensity in different areas within the eye's view (i.e., within its visual "field"). A slowly moving edge of light (or darkness) will affect the rhabdomes located on one side (i.e., right or left) of the lens and in one position (i.e., top or bottom) before it affects the other rhabdomes. The synaptic interactions between photoreceptors, in this case inhibitory interactions, occur on terminal branches of the photoreceptor axons 60–80 μm away from the cell bodies (Figures 19, 21, and 23). These inhibitory interactions will amplify the differences in signals elicited from the photoreceptors by nonuniform illumination of the photoreceptor rhabdomes. Even with a five-cell eye therefore, the *Hermissenda* can discriminate in which direction illumination is greater or what is the direction of a moving shadow. This kind of discrimination will obviously contribute to the animal's ability to move toward a light

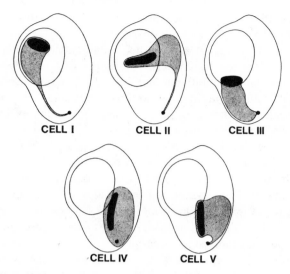

CELL I   CELL II   CELL III

CELL IV   CELL V

Figure 20. Schematic representation of the *Hermissenda* eye showing the general configuration of the five photoreceptors (dark stipple), the rhabdomere (black) beneath the lens, and the course of axons that enter the optic nerve (black dot). Cell IV is situated above cell V, and their rhabdomeres are apposed. (From Stensaas et al., 1969)

source (i.e., show weak positive phototaxis). It will also contribute to its ability to abruptly avoid (i.e., execute an "edge response") a dark border that may arise from a predator's shadow.

Detection of light–dark differences is not only accomplished by the five-cell network within each eye. It is also provided for by the beautifully designed organization of synaptic connections between the visual systems on either side of the animal. Illumination stimulates each eye's cells, which in turn, via synaptic connections, relay this information to a second stage – cells within a structure called the optic ganglion (Figures 24 and 25). The second-stage cells on each side of the animal's "brain" [called the circumesophageal ganglia (Figure 26) because they are ganglia located around the esophagus] send fibers to the opposite side of the brain (Figure 27). These fibers, from the second visual stage of one side, then make inhibitory connections with cells from the second visual stage of the opposite side. What this means is that illumination of one eye will inhibit or reduce the effect of illumination of the opposite side on second-stage cells (Figure 28). This inhibition serves to enhance signaling due to the *difference* between the intensity of light received by the two eyes. Thus, not only are there means of encoding differences in light

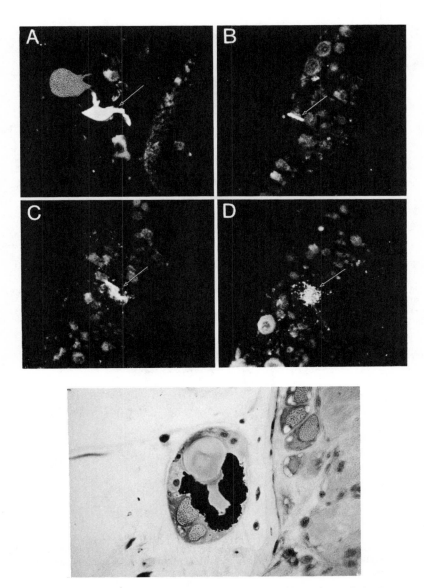

Figure 21. (I) The type B photoreceptor is mapped in micrographs made by Leona Masukawa. The dye Procion yellow was injected into a Type B cell and was transported along its axon to the axon terminals. In the first micrograph (A) the dye fills the cell body (near the gray lens) and the beginning of the axon. In successive thin sections of tissue, (B)–(D), the axon is seen (sectioned obliquely) entering the pleural ganglion and, in the last micrograph, ending in a spray of fine branches. The terminals of the branches form synapses (junctions) with other cells. (From Alkon, 1983) (II) 10-μm section of the *Hermissenda* eye: external epithelium, lens, pigment cup, and two photoreceptors (480 $\times$). (From Alkon, 1976)

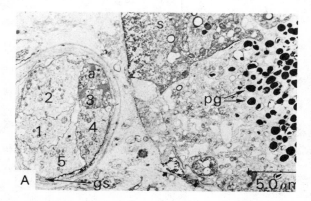

Figure 22A. Electron micrograph showing a HRP-labeled photoreceptor soma (s) and a cross section of optic nerve. The pigment cells contain numerous pigment granules (pg). The optic nerve consists of five axons (1–5) surrounded by a glial sheath (gs). A darkly labeled axon (a) corresponds to the labeled soma. A partially filled area indicated by an arrow (bottom center) may be the result of some leakage from a previous unstable electrode penetration. (From Crow et al., 1979)

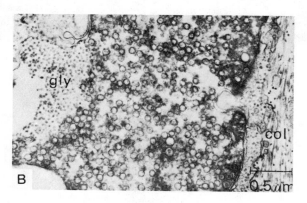

Figure 22B. Electron micrograph at an axon hillock of a HRP-labelled photoreceptor. The area contains numerous vesicles 60–80 nm in diameter, and granular material thought to be glycogen (gly). Outside the stained axon are collagen fibrils (col). A small process on the right of the micrograph invaginates the axon.

intensity within one eye's visual field, there are also means of encoding differences between visual fields. Again, the enhancement (at the second neural stage) of differences between light-elicited signals affecting the two eyes will contribute to the animal's ability to orient and move toward a light source and to avoid dark edges or moving shadows.

In the second neural stage, the cells of the optic ganglion are considered integrating cells or interneurons. They can in turn transmit informa-

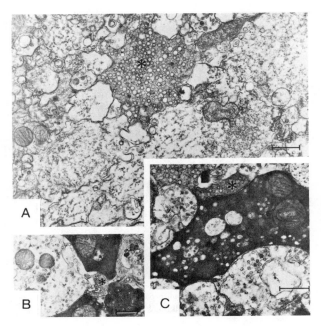

Figure 23. (A) Electron micrograph of a terminal process following injection of a single type B photoreceptor. The terminal process (asterisk) contains numerous clear round vesicles and abuts against several clear processes. Scale bar: 0.5 μm. (B) A moderately labeled terminal process (asterisk) surrounded by three darkly labeled processes following injection of HRP into two type B photoreceptors Scale bar: 0.5 μm. (C) Two terminal processes, probably from two different photoreceptors after injection of two type B photoreceptors. The lightly labeled process (asterisk) contains numerous clear round vesicles and the darkly labeled process contains vesicles and mitochondria. Stain contamination may have obscured some detail in unlabeled areas. Scale bar: 0.5 μm. (From Crow et al., 1979)

tion to other interneurons as well as to motorneurons via synaptic connections. One such motorneuron type, called the MN1 cell (Figure 29), when stimulated by the injection of positive current through an intracellular microelectrode, will cause unilateral contraction of the foot in such a manner as to contribute to turning of the intact animal (Figure 30). The MN1 cell receives excitation (Figure 31), via a known synaptic pathway, from the *Hermissenda* eyes (see the following section). One route taken by signals from the eye is through the second-stage optic ganglion cells to vestibular sensory neurons (called hair cells)) to specific interneurons and thence to the MN1 cells. Another route circumvents the hair cells and begins with the specific interneurons. In each case, these routes for information flow through neuron chains linked by

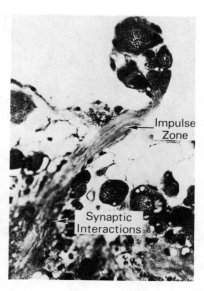

Figure 24. 25-μm thick section stained with toluidine blue showing the optic ganglion (40 μm across) and "optic tract." (From Alkon, 1973a)

synaptic connections provide a neural basis for visually guided orientation behavior and direction of movement.

In addition to the light–dark differences and illumination gradients, diffuse light also has a general arousing effect on the animal. The animals increase their degree and rate of movement even in a uniformly illuminated area. One neural manifestation of this arousal effect is the light-elicited increase of the action potential frequency of particular motorneurons that have a general or widely distributed control of certain kinds of muscle action. For example, the largest neuron in the *Hermissenda* brain (Figure 32) (called the pedal 1 cell) controls movement of featherlike appendages that cover the animal's dorsal surface or back. These appendages, known as ceratae, undergo a wave of movement in response to a burst of action potentials elicited by positive current injection into pedal 1.

Other light-elicited responses of *Hermissenda* involve coordinated control of a number of muscle groups, for instance, those in the animal's foot. As mentioned earlier, the *Hermissenda* foot undergoes a measurable lengthening in response to light onset and an accompanying oriented movement toward a light source. As a result of learning an association between light and rotation, light onset elicits a new behavioral response, foot contraction. This learned response, occurring within

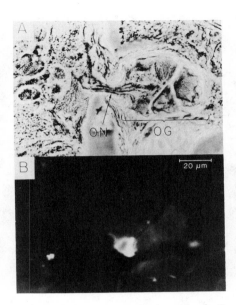

Figure 25. Formaldehyde-induced fluorescence in the optic ganglion (OG). (A) Phase contrast optics showing the ganglion and individual axons within the optic nerve (ON). (B) Fluorescence microscopy of the above. One cell within the optic ganglion has intense greenish fluorescence indicative of catecholamines. Note that the fluorescent material is concentrated in the cytoplasma and in the axon, but not in the nucleus. (From Heldman et al., 1979)

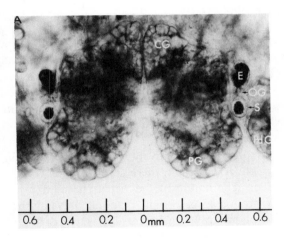

Figure 26. *Hermissenda* circumesophageal nervous system. The photomicrograph is of a living preparation. These statocysts lie between the two central cerebropleural ganglia and the two lateral pedal ganglia. Immediately above the statocysts are the optic ganglia (transparent) and above them are the eyes (darkly pigmented). (From Alkon, 1980)

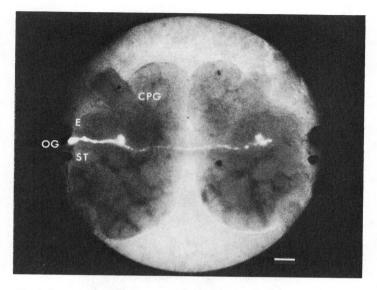

Figure 27. A C optic ganglion cell stained with Lucifer yellow. Lucifer yellow was injected into the cell soma within the left optic ganglion. The axon extends into the contralateral cerebropleural ganglion (CPG) and terminates within that ganglion's neuropil. Branches are visible in both the ipsi- and contralateral CPG. Calibration: 100 μm. (From Tabata and Alkon, 1982)

several seconds of light onset, could involve synaptic connections between the type B photoreceptors and still another group of interneurons found within the ganglia called "cerebropleural." Here there are neurons that receive direct excitatory synaptic input (most likely without any intervening cells) from both the type B cells and the vestibular sensory or "hair" cells (Figure 31). As will be discussed later, these cerebropleural interneurons could be an important convergence site where learning-induced change of visual synaptic input results in the transfer of the unconditioned rotation-elicited foot contraction to the conditioned light-elicited response.

In general, in spite of the relatively small number of cells within the visual pathways of *Hermissenda*, there is a provision for a considerable range of behavior. The nature of the visual stimulation will determine which cells, or groups of cells, will predominate and thereby control which behavior or groups of behaviors. It is clear, for example, that synaptic outputs from the five photoreceptors have distinct effects on different neurons and thus neuronal chains. Conditions that favor stimulation of certain photoreceptors will activate certain pathways and not others and thus produce a particular pattern of behavioral responses.

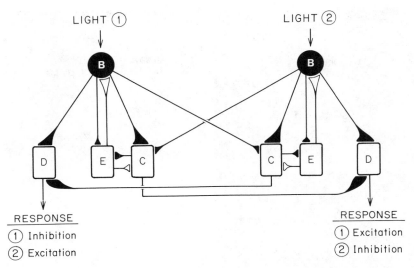

LIGHT ① LIGHT ②

RESPONSE
① Inhibition
② Excitation

RESPONSE
① Excitation
② Inhibition

Figure 28. Visual integration by the *Hermissenda* nervous system. Representative neural elements are included. Light 1 impinges on the left eye, which contains five photoreceptors, here represented by one of the three type B cells. (Similarly, light 2 impinges on the right eye.) Type B impulse activity causes synaptic potentials in 13 of 14 ipsilateral optic ganglion cells, here represented by one D cell, a C (or contralateral) cell, and an E cell. As a result of this neural organization, light 1 causes inhibition, whereas light 2 causes excitation of the left D optic ganglion cell. The converse is true of the right D optic ganglion cell. Filled processes represent inhibition and open, excitatory synaptic interactions. (From Tabata and Alkon, 1982)

Type A photoreceptors are less sensitive to dim illumination than are type B cells. During dim illumination, therefore, type B cell-stimulated pathways may contribute to the weak arousal of the animal and perhaps some capacity to detect light–dark differences. With more intense illumination, activation of the type A cell may lead to visually guided turning movements of the animal and foot lengthening. With very bright light, type A cells retain their previous ability to respond to light, whereas type B photoreceptors actually reverse their light-induced changes of impulse activity (i.e., instead of increasing their impulse activity in response to light, they respond with decreased impulse activity). Thus, with prolonged bright light, with "light-adaptation," type B controlled arousal and light–dark detection may be obscured or eliminated. Behaviorally, the animal might be expected to stop moving in such bright light and thereby show a "preference" for not too high a level of light intensity. In fact, this is actually what intact *Hermissenda* do. They will orient and move toward a light source until they reach the

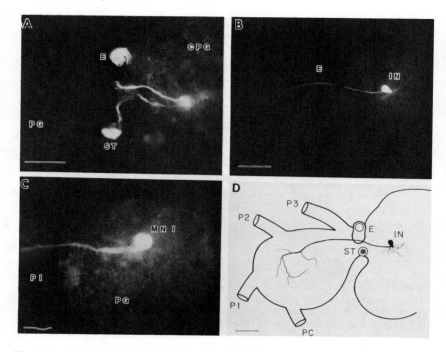

Figure 29. Cells filled iontophoretically with Lucifer yellow. (A) The medial type A photoreceptor and caudal hair cell were stained simultaneously. Processes of both cells terminate at the central area of the cerebropleural ganglion (CPG). (B) The interneuron soma surrounded by some of its axonal branches is located in the same area of the CPG. The axon courses into the pedal ganglion where it terminates in additional fine branches. (C) The MN1 cell is located in the pedal ganglion. The MN1 cell has only one branch going into a P1 nerve. Note the position of LP1 and 2. (D) Drawing of the interneuron. E, eye; ST, statocyst; CPG, cerebropleural ganglion; IN, interneuron; PG, pedal ganglion; P1–P3, pedal nerves 1 to 3; PC, pedal connective. Scale bar: 100 μm. (From Goh and Alkon, 1984)

most intense light within the gradient. They will then avoid the very intense light and tend to remain at its periphery (Figure 33).

How visual stimuli control *Hermissenda* movement will be comprehensively understood only after much further study – particularly of the motor portion of its nervous system. Nevertheless, enough is known about the visual pathways to appreciate that behavioral outcomes depend on the *balance* of electrical activity of the interconnected neural elements within these networks. And how this balance is shifted (say during learning) within a given portion of the network, for instance within the five cell *Hermissenda* eye, can have a profound effect on the

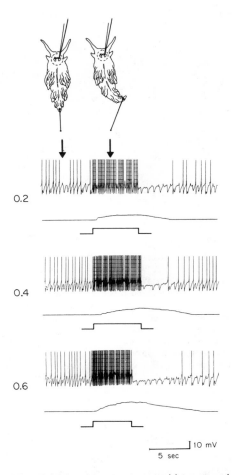

0.2

0.4

0.6

10 mV

5 sec

Figure 30. Motor activity of the foot in response to positive current injection into a MN1 cell. When MN1 impulse activity (upper traces) is elevated by a positive current pulse (nanoamperes indicated by left values), the posterior half of the foot shows turning (lower traces) toward ipsilateral direction (with respect to the stimulated MN1 cell) as shown in schematic figures. Note that the increase of the turning magnitude, as monitored by a stylus attached to a strain gauge, is related directly to the increase of MN1 impulse frequency. (From Goh and Alkon, 1984)

integrated behavioral output elicited. Furthermore, given the clear distribution of visual information along distinct neuronal paths, manifestation of a learned association between stimuli can be expected to be widespread not only within the nervous system but also within the animal's visually

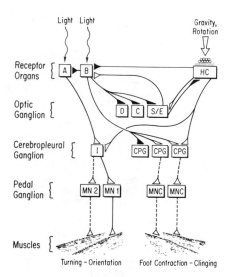

Figure 31. The flow of visual and graviceptive information and their learned interactions may be distributed through identified and predicted neural pathways in such a way that different and even opposite behavioral responses will result. Convergence of information abut the simultaneous occurrence of light and rotation is known to occur at the receptor level, the optic ganglion, and in interneurons found in the cerebropleural ganglion. Pairing-specific increases in the activity of type B photoreceptors may lead to foot contraction in an excitatory pathway through CPG interneurons (Akaike and Alkon, 1980). In a test situation requiring responses to contrast, reduced orientation capabilities may be produced via inhibitory pathways through the medial type A photoreceptors. (From Lederhendler et al., 1986)

controlled behavioral repertoire. Thus, the learning of an association between light and rotation need not be, and is not manifest by, only one measure of *Hermissenda* behavior in response to light. Training with paired light and rotation does, as already mentioned, result in the acquisition of a new response to light. After conditioning, light elicits foot contraction similar to that elicited by rotation (before and after the conditioning). But other light-guided responses reveal conditioning-specific modification as well. The time taken by conditioned animals to move a given distance to reach a light source increases. The ability of the conditioned animals to turn at a light–dark border decreases. Not all light-elicited behaviors are modified by *Hermissenda* conditioning. Whereas the velocity of movement toward a light source (i.e., up a gradient of light intensity) is reduced, velocity of movement in a uni-formly illuminated area is unchanged. Whereas oriented movement in response to moderate light intensities is modified, oriented movement in

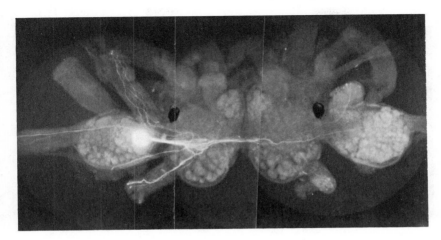

Figure 32. Lucifer yellow stain of LP1. The axon appears to have extensive arborizations in the ipsilateral cerebropleural ganglion before fibers exit through nerves 1, 3, and 4, and cross the midline. Visible from direct microscopic observation were fibers also exiting through the contralateral nerves 1, 3, and 4. LP1 neurons in five preparations stained comparable (magnification 70 ×). (From Jerussi and Alkon, 1981)

response to dim or bright light intensities is not modified. Thus, there is both specificity in the learned association of light and rotation as well as a limited generality. Specificity is demonstrated by the finding that it is only responses to light (and not other stimuli such as food, the pull of gravity, etc.) that are modified, and, moreover, it is only responses to certain types of light stimuli that are modified. Generality is demonstrated by the observation that a number of different light-elicited behaviors are modified as a result of conditioning.

Specificity together with generality are characteristic of associative learning of higher animals (mammals included) as well as of *Hermissenda*. Generality is illustrated, for example, when a rabbit is first conditioned to blink in response to an auditory stimulus (a tone with a particular frequency), it subsequently learns more rapidly to flex its leg in response to the same tone. Certainly our ability to generalize from one learning situation to another cannot be compared to, and is not approximated by, that of *Hermissenda* (or similar creatures such as the land snail *Limax*). Yet it is interesting that a rudimentary form of generalization may be present in the lower animals. This feature of *Hermissenda* learning has some implications as to where the learned information might be stored. To be generalized (i.e., to be distributed to different integrating and motor cells responsible for different behaviors),

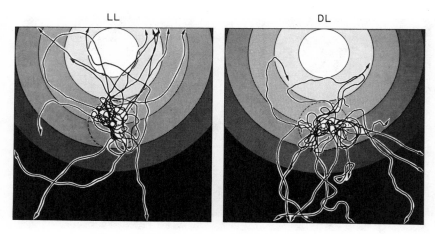

Figure 33. Superimposed tracings of individual *Hermissenda* mucus trails during first 5 min of observation. The center of illumination is the center of the white circle. Individuals were always started in the center of the field. (LL): During the day (0800–1000 h) movement was toward the brighter region of the field but individuals rarely entered the most intense region. (DL): During the night (2000–2200 h) movement was into the darker region of the field. The gradient of illumination is represented by successively shaded zones. Average intensities (in ergs/cm² s) are, respectively: 1.3×10^4 (0–4 cm), 1.0×10^4 (4–8 cm), 5×10^3 (8–12 cm), 1.3×10^3 (12–16 cm), 2.5×10^2 (16–20 cm), 0.2×10^2 (> 20 cm). (From Lederhendler et al., 1980)

it would be stored within cells that send signals in a distributed fashion to these various cells. Those cells (and the behavior they control) that receive a common input may undergo common changes (showing some generality) if that input is modified. Here, *integration of storage sites* for learned information would be accomplished on cells that distribute their information, via synaptic connections, to different neuronal chains necessary for effecting distinct aspects of behavior.

As for the *Hermissenda* visual system, there is a considerable range of behaviors provided by the function of a comparatively few cells within the vestibular system. The animal is negatively geotactic, that is, it orients its movement in a direction opposite to the direction of the earth's gravitational pull. *Hermissenda* will frequently right themselves when turned upside down. Turbulence, shaking movement, or rotation will cause *Hermissenda* to cling to surfaces. All of these behaviors are, to a considerable degree, mediated by two primitive vestibular organs known as statocysts. These cysts contain crystals similar to those of our own middle ear (Figures 34 and 35). The movement of these crystals, due to the animal's movement, gravity, and other forms of acceleration,

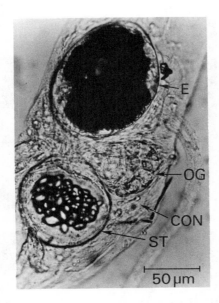

Figure 34. Combined (isolated) sensory structures. E, eye; OG, optic ganglion; ST, statocyst; and CON, capsule of connective tissue. (From Heldman et al., 1979)

stimulates the sensory cells, hair cells, that make up these statocysts. As in the five-cell photoreceptor network in each eye, the 12-cell hair cell network in each statocyst uses inhibitory synaptic interaction to enhance differences between hair cell signals elicited by a gravitational stimulus pattern, such as when the animal is turned upside down. The contrast between excitation of hair cells on the bottom half of the cyst, for instance, and cells on the top half is increased by the inhibitory synaptic interactions. Similarly, contrast between stimulation of hair cells of the right and left statocysts is enhanced by synaptic inhibition between hair cells of the two statocysts.

As mentioned earlier, signals from the statocyst hair cells are, similar to visual signals, in that they are distributed along several different neuronal chains or pathways. Neurons that receive synaptic signals from the hair cells can be found in the eye, the optic ganglia, the cerebral ganglia, and the pedal ganglia. In at least one case, an entire chain has been determined, tracing the vestibular input to the motor output of the system. This particular chain shares important common elements with the visual system and, as has already been described, mediates turning behavior (Figure 31). Thus, the same integrating and motorneurons may implement turning movement, but, depending on the stimuli in the

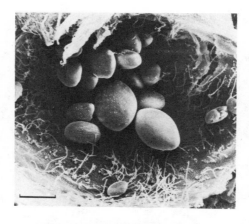

Figure 35. Scanning electron micrograph of the *Hermissenda* statocyst. (A) On the luminal surface of the statocyst, hairs make contact with statoconia. (B) Higher magnification of the above. (From Grossman et al., 1979)

environment, turning may be guided by input from different sensory modalities. The animal may turn to correct its direction of motion so as to be opposite the direction of the earth's pull. Or the animal may turn to correct its direction of motion so as to be toward a light source. When one type of sensory stimulation is of a much greater intensity than another type, it may come to determine the behavioral outcome. An optimally designed nervous system would have pathways that share neuronal elements responsible for turning (thus providing for choice behavior) as well as pathways that exclusively mediate responses to one modality or another of sensory input. Our knowledge of the *Hermissenda* nervous system is in many respects inadequate to answer definitively whether such exclusive pathways also exist in parallel to the one pathway well known to be shared by both the visual and vestibular systems. Enough is clear, however, as will become more evident later in our discussion, to reconstruct how the relationship of visual and vestibular stimuli can be represented within the responses of the networks and how such a representation can be specifically recalled. With critical elements of the pathway for turning shared by the visual and vestibular systems, a locus for a learned transformation of a turning response to one stimulus, light, cannot be in the equally shared elements. Otherwise, the response to the statocyst stimulation will also be modified and the specificity of the learned relationship would be lost. That this specificity is preserved during *Hermissenda* associative learning has important implications for localizing where in the neuronal pathways the learned information is actually stored for later recall. Turning behavior in response to *geotactic stimuli* that is, in response to statocyst stimulation, is not changed by training with paired light and rotation. Turning behavior in response to *light* is changed by the same associative training. Therefore, the learning-induced change, which is recalled, must reside within those neurons sending light-elicited signals to the interneurons and motorneurons that control turning behavior not in the interneurons and motorneurons themselves. Otherwise, hair cell input would also elicit behavior changes as well. These neurons in which learning-induced changes resides, as will be described later, are in the case of the mollusc *Hermissenda*, the photoreceptors themselves. This requirement for *segregation of storage sites* to allow for a specific stimulus or stimulus pattern to recall a memory and to prevent other stimuli from recalling the same memory is a general one. It must be satisfied whether in a rabbit or a snail, a bee or a human. As such, this general requirement sets constraints on where and how associative learning can occur and thus helps narrow the range of possible models or designs for associative learning systems and their function.

Bibliography

Akaike, T., and Alkon, D. L. (1980). Sensory convergence on central visual neurons in *Hermissenda*. *J. Neurophysiol.* 62:185–202.

Alkon, D. L. (1973a). Neural organization of a molluscan visual system. *J. Gen. Physiol.* 61:444–61.

(1973b). Intersensory interactions in *Hermissenda*. *J. Gen. Physiol.* 62:185–202.

(1974). Sensory interactions in the nudibranch mollusc *Hermissenda crassicornis*. *Fed. Proc.* 33:1083–90.

(1975). Neural correlates of associative training in *Hermissenda*. *J. Gen. Physiol.* 65:46–56.

(1976). The economy of photoreceptor function in a primitive nervous system. In *Neural Principles in Vision*, ed. by F. Zettler and R. Weiler, pp. 410–26. Springer-Verlag, New York.

(1980). Cellular analysis of a gastropod (*Hermissenda crassicornis*) model of associative learning. *Biol. Bull.* 159:505–60.

(1983). Learning in a marine snail. *Sci. Am.* 249:70–84.

(1984). Persistent calcium-mediated changes of identified membrane currents as a cause of associative learning. In *Primary Neural Substrates of Learning and Behavioral Change*, ed. by D. L. Alkon and J. Farley, pp. 291–324. Cambridge University Press, New York.

Alkon, D. L., Akaike, T., and Harrigan, J. F. (1978). Interaction of chemosensory, visual and statocyst pathways in *Hermissenda*. *J. Gen. Physiol.* 71:177–94.

Alkon, D. L., and Fuortes, M. G. F. (1972). Responses of photoreceptors in *Hermissenda*. *J. Gen. Physiol.* 60:631–49.

Crow, T., Heldman, E., Hacopian, V., Enos, R., and Alkon, D. L. (1979). Ultrastructure of photoreceptors in the eye of *Hermissenda* labelled with intracellular injections of horseradish peroxidase. *J. Neurocytol.* 8:181–95.

Detwiler, P. B., and Alkon, D. L. (1973). Hair cell interactions in the statocyst of *Hermissenda*. *J. Gen. Physiol.* 62:618–42.

Goh, Y., and Alkon, D. L. (1984). Sensory, interneuronal and motor interactions within the *Hermissenda* visual pathway. *J. Neurophysiol.* 52:156–69.

Grossman, Y., Alkon, D. L., and Heldman, E. (1979). A common origin of voltage noise and generator potentials in statocyst hair cells. *J. Gen. Physiol.* 73:23–48.

Heldman, E., Grossman, Y., Jerussi, T. P., and Alkon, D. L. (1979). Cholinergic features of photoreceptor synapses in *Hermissenda*. *J. Neurophysiol.* 42:153–65.

Jerussi, T. P., and Alkon, D. L. (1981). Ocular and extraocular responses of identifiable neurons in pedal ganglia of *Hermissenda crassicornis*. *J. Neurophysiol.* 46:659–71.

Kandel, E. R. (1976). *Cellular Basis of Behavior: An Introduction to Behavioral Neurobiology*. Freeman, San Francisco.

Kuzirian, A. M., Meyhofer, E., Hill, L., Neary, J. T., and Alkon, D. L. (1986). Autoradiographic assessment of tritiated agmatine as an indicator of physiologic activity in *Hermissenda* visual and vestibular neurons. *J. Neurocytol.* 15:629–643.

Lederhendler, I. I., Barnes, E. S., and Alkon, D. L. (1980). Complex responses to light of the nudibranch *Hermissenda crassicornis* (Gastropoda: Opisthobranchia). *Behav. Neural Biol.* 28:218–30.

Lederhendler, I. I., Gart, S., and Alkon, D. L. (1986). Classical conditioning of *Hermissenda*: origin of a new response. *J. Neurosci.* 6:1325–31.

Stensaas, L. J., Stensaas, S. S., and Trujillo-Cenoz, O. (1969). Some morphological aspects of the visual system of *Hermissenda crassicornis* (Mollusca: Nudibranchia). *J. Ultrastruct. Res.* 27:510.

Tabata, M., and Alkon, D. L. (1982). Positive synaptic feedback in the visual system of the nudibranch mollusc *Hermissenda crassicornis*. *J. Neurophysiol.* 48:174–91.

5

Relating neural change to behavioral change

We use our wiring diagram of the *Hermissenda* visual and vestibular pathways as a map (Figure 36), a map that shows us the route through the nervous system taken by the light and rotational stimuli. With intracellular microelectrodes we monitor the electrical activity of neurons along the route before, during, and after training, and thereby determine how learning alters the routes taken by stimuli through the nervous system. Do signals flow along pathways that were not accessible before learning an association? Or, do already accessible pathways transmit more intense or weaker signals? To answer these questions we must continually relate the electrical activity we record to the behavior of the animal. A measured alteration, if it is involved in the learning process, should increase as the acquisition of the learned behavior increases. Such an alteration should last, if it is critically involved, as long as the learned behavior lasts. And, as the learned association becomes weaker, that is, as it is forgotten, the neural alteration should diminish.

Alterations of signals elicited by the conditioned stimulus, light, during and long after repeated paired presentations of light with the unconditioned stimulus, rotation, occurred at every major stage within the *Hermissenda* visual pathway. As the animal learned to associate light with rotation, different electrical signals were recorded (in response to light alone) from the sensory cells, integrative cells (interneurons), and motorneurons or output cells. These differences in electrical signaling were related to the acquisition (the learning) of the association, the retention of the association (remembering), and the progressive loss of the association (forgetting). In conditioned animals, light elicits a longer depolarization (i.e., a positive voltage change) and more impulses in the type B photoreceptors (Figures 37 and 38). Light elicits less depolarization and fewer impulses in the type A photoreceptors. Integrative cells (interneurons) receive fewer synaptic signals triggered by the type A

55

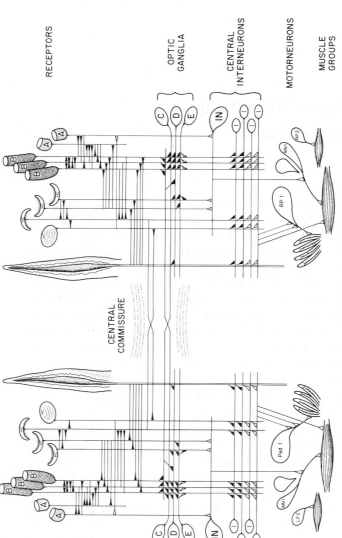

Figure 36. Schematic summary of synaptic interactions in the neural systems of *Hermissenda*. Interactions within and between three sensory pathways are included. The visual pathway begins with the five photoreceptors (two type A and three type B) of each eye. The afferent cells of the vestibular pathway, the 13 hair cells in each statocyst, are labeled HC. The tentacle, represented as an intact structure in the diagram, has chemosensory receptors distributed on its surface. Inhibitory synaptic interactions are indicated by filled endings; excitatory interactions are indicated by open endings. Muscle groups innervated by the motorneurons are pictured at the bottom of the figure. Each interaction represented was established to be reliably present in the adult nervous system by simultaneous pre- and postsynaptic intracellular recording. Not all known interactions are included. (From Alkon, 1982–3)

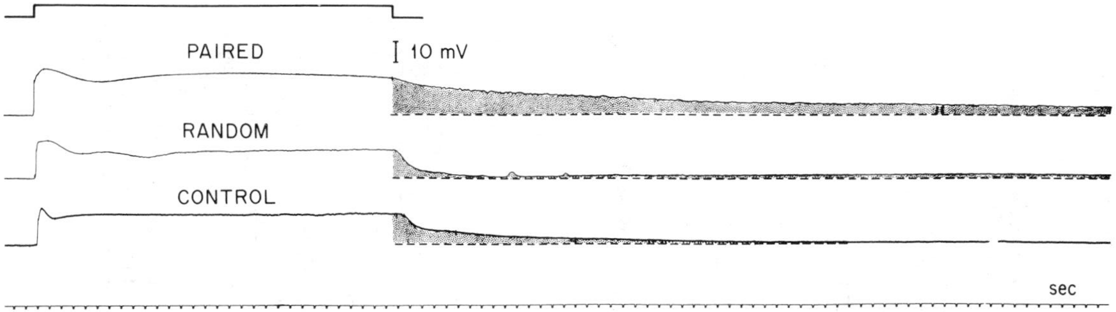

Figure 37. Responses to the first light step of type B photoreceptors from paired, random, and control groups. Shaded areas indicate LLD (long-lasting depolarization) following the light step (monitored by top trace). Note that the paired LLD (i.e., from conditioned animals) is clearly larger than random and control LLD. Impulse and synaptic activity were eliminated by axotomy. (From West et al., 1982).

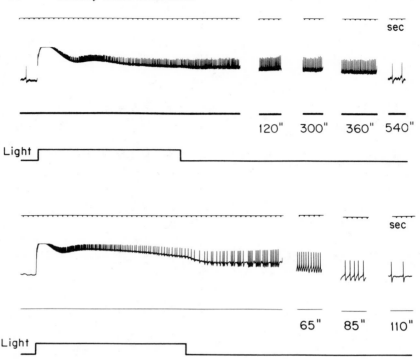

Figure 38. *Conditioned* (upper record). Light responses by a type B photoreceptor from an animal that received paired light–rotation presentations. Recordings were obtained 24 h following training with the preparation oriented vertically (cephalad portion up). The traces present the cell's responses to 30 s of 10^4 ergs/cm^2 s light presented at 2.5-min intervals. Samples of impulse activity at selected times following offset of the light are also depicted. Note the prolonged depolarization of the cell following the response to the light step; it finally returns to initial base line 540 s following light offset. Time scale: 1 s. *Control* (lower record). Light responses by a type B photoreceptor from an animal that received random light–rotation presentations. Recordings were obtained 24 h following training with the preparation oriented vertically (cephalad portion up). The traces present the cell's responses to 30-s steps of 10^4 ergs/cm^2 light, presented at 2.5-min intervals. Samples of impulse activity at selected times following offset of the light are also depicted to illustrate the duration of the cell's depolarization. At 110 s following the light offset, the cell has returned to its initial resting potential level. Time scale: 1 s. (From Farley and Alkon, 1982)

photoreceptor impulses and receive more synaptic signals triggered by the type B photoreceptor impulses. Motorneurons, in turn, receive a different frequency of synaptic signals triggered by interneuronal impulses in response to a light stimulus. In conditioned animals, a lower frequency of motorneuron impulses is elicited by light via synaptic signals received from the photoreceptors and interneurons.

Relating alteration of stimulus-elicited signaling in a *Hermissenda* nervous system to behaviorally manifest learning of an association is a necessary first step for an understanding of how cellular physiology actually brings about the process. For such an understanding, however, we are really confronted by the perennial challenge to the natural observer: discriminating cause from effect. At the outset, when the training experience begins, a particular pattern of sensory stimuli, and the repeated presentation of that pattern, are causes of particular patterns of electrical signaling within the nervous system. Repetition of the stimulus pattern *causes* the electrical signal pattern to change. Later in the process, when a stimulus, part of the original pattern, is subsequently presented after training, it causes a new pattern of electrical signals and a new behavior – the learned behavior that expresses the association of light and rotation. But the subsequent presentation of a stimulus is not the only *cause* of the new behavior. Within the nervous system, something has changed to *cause* a new set of electrical signals during subsequent presentation of a conditioned stimulus (here, light). That something that has changed, *together* with the stimulus, *causes* the new signals. That changed something is, in effect, storing information about the prior stimulus patterns experienced by the animal during training. The stored information is retrieved by subsequently presented conditioned stimuli.

Given that changes of the photoreceptors', interneurons', and motorneurons' electrical signals can be related to associatively learned behavior, can we assume that all of these cells are sites for storing the learned information? Not by any means. Impulse activity of cells can change, as already mentioned, because of change in synaptic signals received by those cells. These cells may be storing no information at all but, by their impulse frequency, simply be reflecting or expressing information stored elsewhere that influences synaptic input to many cells. The electrical signals of these cells that do not store information are a consequence of, an effect of, rather than a cause of, information storage during learning. Changes of electrical signals that are effects of, rather than causes of, information storage can be considered as secondary, whereas primary changes of signaling emanate from cells that are actual storage sites.

A large body of experimental evidence indicates, for example, that type B photoreceptors are sites for storing the learned association of light and rotation. Unlike the primary changes of type B signals during learning, secondary changes can be recorded from interneurons and motorneurons that receive type B synaptic signals.

What kinds of evidence can be brought to bear to distinguish a primary from a secondary learning-induced change? Two types of evidence, already discussed, are necessary but not sufficient: (1) close correlation of the learning-induced neural change with the learning-induced behavioral change; (2) organization of synaptic interactions such that the proposed primary neural change can account for other secondary neural changes. The type B photoreceptors are at the input end of the visual pathway. Changes of type B electrical signals that result in changes of synaptic output must affect those cells that receive the synaptic output. The organization of synaptic interactions in the *Hermissenda* visual system is such that type B cell synaptic output affects a turning pathway (consisting of interneurons and motorneurons), central pleural neurons that could influence contraction of the foot, and optical ganglion cells. Thus, there are three separate routes of information flow; all of the visually responsive interneurons and motorneurons studied thus far in the *Hermissenda* nervous system receive directly or indirectly synaptic input from the type B cells. By virtue of the direction of information flow, therefore, when the type B cells undergo learning-induced modification, they are very likely to cause secondary change throughout the known visual pathways.

To further establish the type B cell as a storage site, we might first consider an idealized experiment. If information is stored within a neuron, that information should remain even after all synaptic interaction with that neuron has been eliminated, that is, the stored information should be independent of communication between the storage site and other loci in the nervous system. Ideally, it should be possible to reach into the nervous system and pluck out the storage-site neuron and then to identify learning-induced changes in this isolated neuron. The biological record of the learned association, or a part of it, would be isolated within the neuron.

As an approximation of this experimental ideal, type B cells were isolated from the nervous system by severing their axons. Because all synaptic interactions occur on the distal portion of the axons (Figures 19 and 83), they are removed by a cut near the axon's point of origin near the base of the cell body or soma. Electrophysiologic measurements of type B cell bodies isolated in this manner from associatively conditioned animals as well as a variety of controls (such as those trained with randomly occurring light and rotation) demonstrated conditioning-specific changes. Membrane properties changed by conditioning persisted in the isolated type B soma (Figures 37 and 39). The learning-induced changes observable for the membranes of isolated type B cell

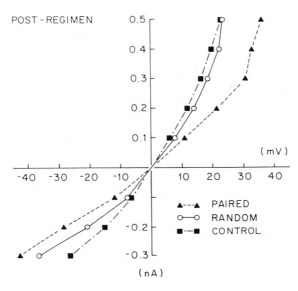

Figure 39. Current–voltage relations of type B photoreceptors. One or two days after the training regimen, values were obtained by measuring the steady-state voltage changes produced by current pulses through an intracellular microelectrode. The paired cells showed significantly greater voltage changes, particularly for positive pulses. (From West et al., 1982)

bodies were of the type and magnitude necessary to account for other electrophysiologic changes observed for nonisolated cells (Figure 38), cells with intact axons and synaptic interactions. The isolated learning-induced changes were responsible for much of the learning-induced modification of the type B cell's response to the conditioned stimulus, light, and, in turn, for secondary changes of the visual pathway neurons that receive synaptic signals from the type B cells.

Additional evidence of a storage site involves artificial production or synthesis of learning-induced neural and behavioral change. Again, an experimental ideal motivates experimental design. We begin with the hypothesis that during acquisition of a learned association, signal patterns along particular pathways (i.e., along certain neuronal chains whose links are joined by synaptic interactions) produce neuronal changes that record for later recall the repeated occurrence, with a fixed temporal relationship, of distinct sensory stimuli (such as light and rotation). Ideally, we should be able to remove (by dissection) the nervous system together with the sensory organs generating signals in response to the training stimuli. Ideally, repeated presentations of the same stimulus pattern, used for training the animal to the isolated nervous system, will result in the same neuronal changes that were correlated with learning behavior of intact animals. Ideally, we should be able to insert microelectrodes into neurons within the pathways of interest and inject microcurrents to produce the same electrical signals that occur during associative training and thereby produce the same neuronal changes that were observed during and after the learning. And finally, ideally, we should be able to insert microelectrodes into neurons of the living animal and with the appropriate current injections produce not only the learning-induced neuronal changes but also the learning-induced behavioral change.

All of these experimental ideals have proved possible to approximate for *Hermissenda* associative learning. Repeated presentation of light and rotation (just as occurs during training of intact animals) to the isolated nervous system (which includes the eyes and statocysts) causes the same type B cell changes that were correlated with the acquisition and retention of learned behavior in the intact animal. Injection of current through microelectrodes inserted into type B cells and/or in particular statocyst hair cells and optic ganglion cells could be timed with the onset of light so as to produce the same learning-induced modifications of the type B cell. And, even in type B cell bodies or somata that had been isolated (by axotomy) from all synaptic interaction, critically timed current injection, when paired with light stimuli, could produce type B changes similar to those observed with learning.

Finally, penetration of type B cells was accomplished with microelectrodes in living *Hermissenda*. Repeated pairing of current injection with light stimuli resulted in the learning-induced modification of type B cell properties and, on days after removing the microelectrodes, behavioral change was observed to be similar to that which was associatively learned. Thus, production of the type B cell changes in the living animal was sufficient to produce the learning-induced behavioral change.

These experiments taken together provide conclusive evidence that the type B cell is one important locus for storing information concerning the temporal association of light and rotation. The stored "information" is accessed, is recalled, when the animal subsequently encounters a light stimulus in the absence of rotation. The "information" accessed is that the light stimulus has in the past been followed and accompanied by rotation. That the information is recalled is behaviorally manifest by a new response to the light stimulus – a light response that now is similar to the previous response to rotation.

These experiments do not provide conclusive evidence that the type B cell is the *only* locus for storing information necessary for learning the association between light and rotation. In fact, other experiments indicate that type A photoreceptors undergo conditioning-specific changes that are complementary to those of the type B cells. Type A cells becomes less responsive, whereas type B cells become more responsive to light stimuli as a result of conditioning. Therefore, even in the relatively simple nervous system of *Hermissenda*, learning is probably recorded by a small set of neurons. We might anticipate that, in much more complex nervous systems, the set involved in recording an association contains a vastly increased number of cells – a number orders of magnitude larger than that in *Hermissenda*. With such a number the resolution, specificity, and generalizability of the association should also be vastly increased. Thus, the phenomenology of the learning behavior, as well as that of the underlying neural systems' physiology in a vertebrate, can be expected to include features absent for that of *Hermissenda*, There may, however, be common principles that apply and that ultimately have expression in common learning-specific transformations of membrane and molecular properties of neurons that are in fact primary loci of change – loci that store the learned information. As will become apparent, a remarkable similarity has now been observed in learning-specific changes of particular hippocampal neurons and those for the type B cells, giving some encouragement to our expectation that on membrane and molecular levels mechanisms for recording learned information have been conserved for a considerable interval of evolutionary time.

Bibliography

Alkon, D. L. (1975). Neural correlates of associative training in *Hermissenda*. *J. Gen. Physiol.* 65:46–56.

(1982–3). Regenerative changes of voltage-dependent Ca^{2+} and K^+ currents encode a learned stimulus association. *J. Physiol. Paris* 78:700–6.

Alkon, D. L., Lederhendler, I., and Shoukimas, J. J. (1982). Primary changes of membrane currents during retention of associative learning. *Science* 215:693–5.

Alkon, D. L., Sakakibara, M., Forman, R., Harrigan, J., Lederhendler, I., and Farley, J. (1985). Reduction of two voltage-dependent K^+ currents mediates retention of a learned association. *Behav. Neural Biol.* 44:278–300.

Berger, T. W., and Thompson, R. F. (1978a). Neuronal plasticity in the limbic system during classical conditioning of the rabbit nictitating membrane response. I. The hippocampi. *Brain Res.* 145:323–46.

(1978b). Neuronal plasticity in the limbic system during classical conditioning of the rabbit nictitating membrane response. II. Septum and mammillary bodies. *Brain Res.* 156:293–314.

Buchwald, J. S., Halas, E. S., and Schramm, S. (1966). Changes in cortical and subcortical unit activity during conditioning in chronic cats. *Physiol. Behav.* I:11–22.

Camardo, J. S., Siegelbaum, A. S., and Kandel, E. R. (1984). Cellular and molecular correlates of sensitization in *Aplysia* and their implications for associative learning. In *Primary Neural Substrates of Learning and Behavioral Change*, ed. by D. L. Alkon and J. Farley, pp. 185–203. Cambridge University Press.

Cohen, D. H. (1984). Identification of vertebrate neurons modified during learning: analysis of sensory pathways. In *Primary Neural Substrates of Learning and Behavioral Change*, ed. by D. L. Alkon and J. Farley, pp. 129–54. Cambridge University Press.

Crow, T. J., and Alkon, D. L. (1980). Associative behavioral modification in *Hermissenda*: cellular correlates. *Science* 209:412–14.

Farley, J., and Alkon, D. L. (1982). Associative neural and behavioral change in *Hermissenda*: consequences of nervous system orientation for light and pairing specificity. *J. Neurophysiol.* 48:785–807.

Farley, J., Richards, W. G., Ling, L. J., Liman, E., and Alkon, D. L. (1983). Membrane changes in a single photoreceptor cause associative learning in *Hermissenda*. *Science* 221:1201–3.

Goh, Y., Lederhendler, I., and Alkon, D. L. (1985). Input and output changes of an identified neural pathway are correlated with associative learning in *Hermissenda*. *J. Neurosci.* 5:536–43.

Kanz, J. E., Eberley, L. B., Cobbs, J. S., and Pinsker, H. M. (1979). Neuronal correlates of siphon withdrawal in freely behaving *Aplysia*. *J. Neurophysiol.* 42:1538–56.

Richards, W. G., Farley, J., and Alkon, D. L. (1984). Extinction of associative learning in *Hermissenda*: behavior and neural correlates. *Behav. Brain Res.* 14:161–70.

Thompson, R. F., Barchas, J. D., Clark, G. A., Donegan, N., Kettner, R. E., Lavond, D. G., Madden, J., IV, Mauk, M. E., and McCormick, D. A. (1984). Neuronal substrates of associative learning in the mammalian brain. In *Primary Neural Substrates of Learning and Behavioral Change*, ed. by D. L. Alkon and J. Farley, pp. 71–99. Cambridge University Press.

West, A., Barnes, E. S., and Alkon, D. L. (1982). Primary changes of voltage responses during retention of associative learning. *J. Neurophysiol.* 48:1243–55.

Woody, C. D., and Engel, Jr., J. (1972). Changes in unit activity and thresholds to electrical microstimulation at coronal–pericruciate cortex of cat with classical conditioning of different facial movements. *J. Neurophysiol.* 35:230–41.

6
Design of neural systems

The neural organization, that is, the aggregate of known synaptic interactions between neurons of the visual pathway, is consistent with a primary or causal role for conditioning-specific type B cell changes. Similarly, for the same learned association, the synaptic interactions involving neurons of the vestibular (or statocyst) pathway are *in*consistent with a causal role for conditioning-specific changes within a number of other neurons. It is known that statocyst sensory cells send signals to, that is, release synaptic messages onto, the same interneurons that also receive such signals from visual sensory cells (i.e., the photoreceptors). These interneurons include those involved in turning movements of the animal (as well as others within the cerebropleural ganglion). It is also known that turning movement controlled by the statocyst cells' responses to the animal's position in space is not altered by the learned association, that is, the learned association is stimulus-specific. This stimulus specificity of the learned behavioral response would be lost were conditioning-induced changes to alter the way interneurons (or motorneurons) receive signals from *both* the eye and statocyst. Changes on the type B photoreceptors, however, will not alter signals received from the statocyst hair cells in response to gravitational stimuli, thereby preserving the stimulus specificity of the learned behavior. Thus, it is reasonable to expect that interneurons and motorneurons in the *Hermissenda* pathway involved in the animal's turning movement undergo only secondary conditioning-induced changes. After the association between light and rotation is learned, subsequent presentations of the conditioned stimulus, light, will elicit a new signal pattern from interneurons and motorneurons in the turning pathway, whereas statocyst stimulation will not elicit a new signal pattern from these interneurons and motorneurons. In fact, it has been demonstrated that when the type B cell change is known for a conditioned *Hermissenda*, it is possible to predict with considerable accuracy a change of a motorneuron's (in the visual pathway) response to light.

Interneurons that receive both visual and statocyst input could, how-ever, store conditioning-specific information in a way that would pre-serve stimulus specificity. This way involves compartmentalizing the learning-induced change on the structure of the interneuron. A suffi-ciently compartmentalized change could conceivably alter the signals received from the eye and leave the signals received from the statocyst unaltered. Compartmentalization of this kind would require that the membrane properties of one compartment did not significantly de-termine either the general excitability of the cell or the reception of signals in the other compartment. Thus, learning-induced change of an interneuron common to both the visual and statocyst pathways could not involve the cell body or the major axonal trunk, and still preserve stimulus specificity. Stimulus specific learning, however, could involve localized changes of small branches that receive visual input and not involve small branches that receive statocyst input.

In contrast to a compartmentalized structural locus for learning-induced changes of shared interneurons, a noncompartmentalized locus (one affecting general excitability) could store learned information in an unshared cell. The type B photoreceptor is directly in the pathway of visual signals flowing through the nervous system but not of statocyst signals. The excitability of the type B cell is changed by conditioning thereby clearly influencing visually elicited signals received by interneu-rons and motorneurons. And although it is true that the type B cell also receives and sends synaptic signals from and to the statocyst, the statocyst-generated signals (in response to gravitational stimuli) to inter-neurons are not greatly determined by the type B statocyst interaction.

Suggested principles of design

These considerations suggest some general design constraints for neural systems responsible for mediating a learned association. The stimuli to be associated must affect signaling in some common cells, that is, the associated stimuli must "converge" within the neural system. As will be discussed later, convergence is crucial for bringing about the learned association. Yet at the same time within the neural system, there must be sufficient separation of cells sensitive to either of the training stimuli so that learning-induced transformations can be stimulus-specific and can affect the response to one stimulus but not another stimulus. And, within the neural design, there must be some communication between the loci of convergence and the loci responsive separately to the conditioned and

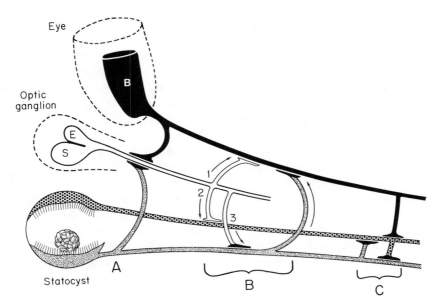

Figure 40. Intersensory integration by the *Hermissenda* nervous system. (A) Convergence of synaptic inhibition from the type B and caudal hair cells onto the E cell. (B) Positive synaptic feedback onto the type B photoreceptor. 1, direct synaptic excitation; 2, indirect excitation: E excites the cephalic hair cell that inhibits the caudal hair cell and thus disinhibits the type B cell; 3, indirect excitation: E inhibits the caudal hair cell and thus disinhibits the type B cell; 4, indirect excitation: the B cell inhibits the C cell, and thus disinhibits the E cell; C cell effects are not illustrated. (C) Intra- and intersensory inhibition. Cephalic and caudal hair cells are mutually inhibitory. The type B cell inhibits mainly the cephalic hair cell. All filled endings indicate inhibitory synapses; open endings indicate excitatory synapses. (From Tabata and Alkon, 1982)

unconditioned stimuli (Figure 31). This communication will involve a direction-of-information flow opposite to that of input (sensory cells) to output (motor cells). If the input-to-output direction is considered as "forward," then the opposite direction of information flow can be considered as "backward" or "feedback" (Figure 40).

To provide sites for convergence of signals to be associated (sites of stimulus-specific learning-induced neural and behavioral changes, and the "feedback" for interaction between loci of convergence and loci responsive separately to the conditioned and unconditioned stimuli), the neural system also must have within its design considerable parallel flow of information or pathways that "diverge" from an input source. Light elicits photoreceptor signals that can send messages to optic ganglion cells that are specialized for contrast detection, optic ganglion cells that receive statocyst input (i.e., are convergence sites), interneurons involved in turning, and still other interneurons in the cerebropleural ganglion. Thus, the visual information diverges from its entry point (the photoreceptors) into the nervous system. As part of this divergence, visual signals from the photoreceptors converge with signals from the statocyst hair cells on common interneurons, which in turn control motor output (Figure 31). Visual signals from the photoreceptors also converge with signals from statocyst hair cells on other common interneurons, particularly optic ganglion cells, called E cells, that may not control motor output but send signals back to the photoreceptor – feedback critically important for modifying the photoreceptor during training with paired light and rotation (Figure 40). The details of the *Hermissenda* neural system's wiring, which provide for divergence, convergence, and feedback, will emerge more clearly later as we explore how the learning of an association is actually accomplished on a cellular level. These details will also suggest parallels that would help to explain related conditioning-specific cellular changes observed in rabbits, and ultimately, perhaps, ourselves as well. For now it will suffice to say that the neural systems allow for *both* mixing of discrete stimuli (at one type of convergence point) as well as for unimpeded separate flow of signals elicited by the stimuli through another type of convergence point and through nonconvergence points to additional pathway elements. These parallel pathways are necessary in order to have both reflexly executed behavioral responses and the potential for modifying other responses in the course of associative training.

Bibliography

Alkon, D. L. (1974). Sensory interactions in the nudibranch mollusc *Hermissenda crassicornis*. *Fed. Proc.* 33:1083–90.

Goh, Y., and Alkon, D. L. (1984). Sensory, interneuronal and motor interactions within the *Hermissenda* visual pathway. *J. Neurophysiol.* 52:156–69.

Lederhendler, I., Gart, S., and Alkon, D. L. (1986). Classical conditioning of *Hermissenda*: origin of a new response. *J. Neurosci.* 6:1325–31.

Tabata, M., and Alkon, D. L. (1982). Positive synaptic feedback in the visual system of the nudibranch mollusc *Hermissenda crassicornis*. *J. Neurophysiol.* 48:174–91.

7

The language of the nervous system

We might regard impulses as providing a voice for the language of the nervous system. Impulses are large changes of voltage across the cell membrane (Figure 41). The magnitude of these voltage changes tends to be quite uniform, whereas their frequency undergoes an enormous range of variation. And so it is that the different frequencies of impulses provide a sort of alphabet for the neural language. Intervals of impulse frequencies can be arranged in different patterns so as to comprise messages or words. The true "message" or linguistic representation emerges, however, not from impulse frequencies of one cell, but from hosts or systems of cells.

Another characteristic of impulses is their ability to travel great distances along the structure of a neuron without becoming smaller. This "propagation" of the impulse occurs with great speed so that impulses are very efficient at carrying messages from one portion of a neuron to another and thus often from one part of the nervous system to another.

One impulse occurs when the voltage across the cell membrane reaches a critically positive level called a "threshold" (comparing the electrical potential inside the cell to the potential outside) (Figure 42). The potential inside the cell is quite negative in comparison to the potential outside when the cell is unstimulated or "at rest." When the potential inside the cell becomes less negative in comparison to that outside, the membrane potential across the cell membrane is said to become more positive. This is a positive voltage change. The cell is stimulated by synaptic messages that can be excitatory or inhibitory. These messages arise as the result of impulses releasing chemicals from neighboring cells (i.e., via a "chemical synapse"). Neighboring cells can also be connected to a neuron, and thereby influence its membrane potential through an electrical "bridge," that is, voltage changes spread directly between cells in the absence of chemical release at loci of structural proximity, called "electrical synapses." An excitatory synaptic message (Figures 86 and

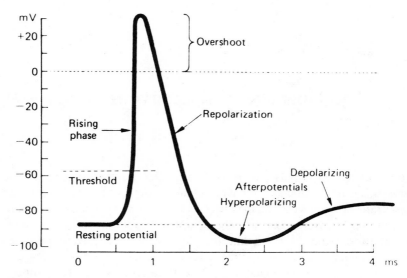

Figure 41. Phases of the impulse or action potential. The time course of a nerve action potential is diagrammed. (From Schmidt et al., 1978)

87) is associated with a positive voltage change (an excitatory potential) that brings the membrane potential to a level closer to where an impulse is triggered. Once this threshold is reached, the voltage changes in an "explosive" or regenerative manner – a sequence of cellular events occurs that makes the impulse a somewhat inevitable consequence of the initial voltage change to threshold. An inhibitory synaptic message (Figures 19 and 85) is associated with a negative voltage change (an inhibitory potential) that brings the membrane potential to a level away from that necessary to trigger an impulse.

A neuron can also be stimulated, if it is a sensory receptor, by a natural stimulus. The type B photoreceptor becomes more depolarized, that is, the voltage across its membrane (inside vs. outside) becomes more positive, when light affects a specialized cellular compartment known as a rhabdome. The rhabdome contains molecules, called rhodopsin, which when exposed to light initiate a sequence of chemical steps that lead to a positive change of membrane potential (or voltage). Light, through its effect on the rhabdome, also stimulates the type B photoreceptor. Gravity, through its effect on hairlike structures of the *Hermissenda* vestibular organ (the statocyst) can stimulate the vestibular sensory receptors called "hair" cells.

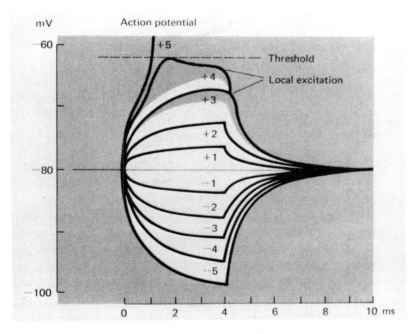

Figure 42. Electrotonic potentials and local responses. Current pulses (4 ms in duration) of relative amplitudes 1, 2, 3, 4, and 5 in the hyperpolarizing direction produce proportional electrotonic potentials. With depolarizing currents of amplitudes 1 and 2, the potentials are mirror images of those with hyperpolarizing currents. The depolarizations produced by current amplitudes 3 and 4 exceed those of electrotonic potentials at levels beyond -70 mV, by amounts indicated by the areas below the curves. The active or nonlinear depolarization, in excess of electrotonus, is called local excitation. The depolarizing current of amplitude 5 produces a depolarization that passes the threshold and triggers an action potential. (From Schmidt et al., 1978)

The depolarizing shift of potential caused by sensory stimuli such as light and rotation are called generator potentials. Generator potentials arising in one compartment can spread passively to another compartment, such as a particular region of the axon where impulses are triggered. The membrane of this axonal region is such that a depolarizing shift, whether it be due to the spread of a generator potential or to the spread (from a synaptic locus) of a synaptic potential, will elicit one or more impulses. The number of impulses elicited is proportional to the magnitude of the positive shift of membrane potential. A current injection of $+0.1$–0.2 nA, for example, will trigger 1–2 type B impulses during 1 s, whereas 0.4–0.5 nA will trigger 8–10 impulses. The brighter the light stimulus, the greater is the positive shift (Figure 43) and the higher is the frequency of impulses elicited (Figures 79–81). A bright

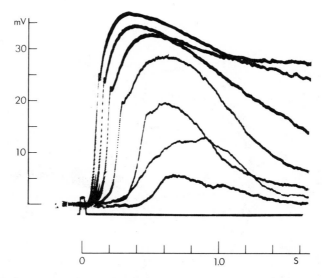

Figure 43. Responses of a photoreceptor with cut axon. Responses are to flashes (indicated by the bottom trace) of a quartz–iodide light source (2×10^5 ergs/cm² s) attenuated by neutral density filters as follows (increasing intensity): 4.5 O.D., 4.2, 3.6, 3.0, 2.4, 1.0, and 1.2. (From Alkon, 1976)

light (during a sustained stimulus such as that shown in Figure 81) may elicit 20–30 type B impulses in 1 s and 2–4 type A impulses. A dim light may elicit no type A impulses but 5–10 type B impulses in 1 s.

Bibliography

Akaike, T., and Alkon, D. L. (1980). Sensory convergence on central visual neurons in *Hermissenda*. *J. Neurophysiol*. 44:501–13.

Alkon, D. L. (1973). Neural organization of a molluscan visual system, *J. Gen. Physiol*. 61:444–61.

(1976). The economy of photoreceptor function in a primitive nervous system. In *Neural Principles in Vision*, ed. by F. Zettler and R. Weiler, pp. 410–26. Springer-Verlag, New York.

Alkon, D. L., and Fuortes, M. G. F. (1972). Responses of photoreceptors in *Hermissenda*. *J. Gen. Physiol*. 60:631–49.

Alkon, D. L., and Grossman, Y. (1978). Long-lasting depolarization and hyperpolarization in eye of *Hermissenda*. *J. Neurophysiol*. 41:1328–42.

Schmidt, R. F. (1978). *Fundamentals of Neurophysiology*. Springer-Verlag, New York.

Tabata, M., and Alkon, D. L. (1972). Positive synaptic feedback in the visual system of the nudibranch mollusc *Hermissenda crassicornis*. *J. Neurophysiol*. 48:174–91.

8

The nature of learning-induced
neural change

In conditioned animals the magnitude of the depolarizing generator potential and the frequency of impulses elicited by a constant light stimulus increase during learning and remain increased while the learning is retained by the animal (Figures 37 and 38). Because the number of impulses elicited by a given light stimulus increases, the type B cell can be considered as more "excitable." This conditioning-induced increase of type B excitability is not restricted to a light stimulus. A positive voltage change across the type B membrane also results from injection of positive current through a microelectrode inserted into the type B cell. A given current injected in this way causes more impulses of type B cells from conditioned, as compared to various control, animals (Figure 39).

If we consider that the type B cell membrane is a simple electrical resistor, we can apply Ohm's law, which states that the potential (or voltage), E, is equal to the magnitude of the electrical resistance, R, multiplied by the magnitude of the current, I, flowing (or injected) through that resistor (i.e., $E = IR$). If, with conditioning, the magnitude of the voltage shift, E, is greater for a constant injection of current, I, then by Ohm's law, (i.e., to satisfy the equation $E = IR$), the electrical resistance of the conditioned type B cell must have increased. A conditioning-specific increase of type B resistance has been measured in intact cells as well as cell bodies that have been isolated by axotomy from all synaptic interaction as well as an impulse-generating membrane. (After axotomy, the axonal ends most likely seal over preventing leakage of intracellular contents.) Thus, the electrical resistance *of the cell body* membrane has increased with conditioning. Any current flowing across the resistance of the cell body will cause a greater voltage change. Light-induced current, originating at the rhabdome, will cause a larger voltage change and thus a large depolarizing generator potential across the type B cell body membrane. A larger conditioning-specific depolarization will spread to the axon and elicit more impulses in a constant time interval. More impulses will spread into the fine terminal branches of the type B cell and there release more chemical messengers at synaptic junctions. A conditioning-specific increase of synaptic transmitter released will initiate a chain reaction ultimately leading to a change of impulses within motorneurons (Figures 36 and 84) that control muscular contraction involved in the animal's behavioral response to light.

And so it is the increased type B cell resistance that is really responsible for a larger light-induced generator potential, more impulses, and a greater release of a synaptic messenger (or a neurotransmitter). It is this increased resistance that persists in the absence of all synaptic interaction and, in fact, all contact with the rest of the nervous system or any other part of the animal. It is this increased resistance that is intrinsic to the type B cell membrane and that stores information about the prior temporal association of light and rotation. But what is increased resistance itself really? The word resistance has an electrical connotation similar to its connotation in other contexts. The type B cell membrane offers resistance to the flow of electrical current. The current flows across the type B membrane through microscopic pores or channels (Figure 44). The larger the open channels and/or the more numerous the open channels within a membrane (as in the type B cell body), the easier it is for current to flow across that membrane. The smaller and fewer the open channels, that is, the more difficult the current flow, the higher the resistance to current flow and the larger the voltage change produced by that current flow.

Electric current is the movement of charged particles. The ions (molecules with electric charge) that flow across biological membranes are well known. They include sodium, chloride, potassium, and calcium. Channels within biological membranes are specialized to permit passage of some ions but not others. Thus, there can be separate channels for sodium, potassium, and calcium ions. And there can even be different channels for the same ion. For instance, certain potassium channels open rapidly but close rapidly as well, whereas others open slowly and may require an activator (such as calcium) inside the cell.

When a cell is unstimulated (i.e., when it is "at rest"), resistance to current flow (and thus the majority of ion channels) across biological membranes (including that of the type B cell) is largely determined by potassium channels. Activation of these potassium channels frequently depends on the level of potential across the cell membrane. When there is a positive shift of membrane potential (inside vs. outside), a certain number of potassium channels open, that is, the channels are "voltage

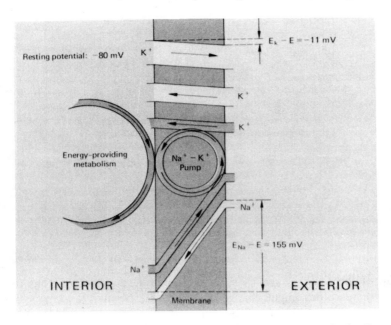

Figure 44. Passive and active movements of ions through the membrane. In the diagram the width of the channels for various ionic currents indicates the magnitude of the current, and the slope of the channels indicates the driving force behind the ionic current. Because of the Na^+-K^+ pump, sodium and potassium currents occur against the direction of the driving force. (From Eccles, 1957)

dependent." This can also be true for sodium and calcium channels. In the type B soma membrane, the potassium and calcium channels (but not the sodium channels) are opened by positive shifts of membrane potential, whereas the sodium channels (at the rhabdome) are opened by a light-initiated sequence of biochemical steps.

The number of ions flowing through distinct channels can be measured (using a technique called "voltage-clamp" as in Figure 45) as distinct ionic currents. When positive ions (e.g. K^+) move from the inside to the outside of the cell it is called "outward" and is represented by an upward movement of the current trace (Figure 46). When positive ions move from the outside to the inside of the cell it is called "inward" and is represented by a downward movement at the current trace (Figure 48). So there are potassium currents, calcium currents, etc., each with defining characteristics. The total complement of ionic currents across a biological membrane such as that of the type B cell is genetically determined. Previously, we discussed how it was necessary to arrive at a

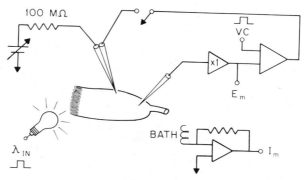

Figure 45. Schematic of the voltage-clamp apparatus. The circuit is a conventional, two-microelectrode voltage clamp with current measured by a virtual ground amplifier. For iontophoresis, a double-barreled micropipette supplied clamp current from one barrel and iontophoretic current out of the other barrel via the dc voltage source and 100-MΩ dropping resistor. (From Alkon et al., 1982)

map of the routes taken by signals through a system of neurons to see how these routes might change with learning. Similarly, we must know the routes taken by ions across the type B membrane to see how these routes might change with learning and by such changes cause the increased resistance that was observed.

There are several major ionic currents that flow across the type B soma membrane:

1. A K^+ current that flows from the inside of the cell to the outside (and is thus called "outward") and activates and inactivates rapidly. This current is activated by positive shifts of membrane potential and is called "I_A" (Figure 46).
2. An outward K^+ current that activates and inactivates slowly and is also activated by positive shifts of membrane potential. This current occurs when the level of Ca^{2+} inside the cell rises and is thus called "$I_{Ca^{2+}-K^+}$," or a calcium-dependent K^+ current (Figure 47).
3. A Ca^{2+} current that flows from the outside of the cell to the inside (and is thus called "inward") and activates rapidly but (unlike the K^+ currents) stays activated – it shows no inactivation. This sustained inward Ca^{2+} current is called "$I_{Ca^{2+}}$" and also arises in response to positive shifts of membrane potential (Figures 48 and 49).
4. An inward Na^+ current that activates rapidly but also inactivates rapidly. The Na^+ current, called "I_{Na^+}," is not activated by positive shifts of membrane potential, but instead by a light-initiated sequence of biochemical steps (Figure 50).

Once these ionic currents have been quantified for untrained animals, the effect of conditioning with paired light and rotation can be assessed.

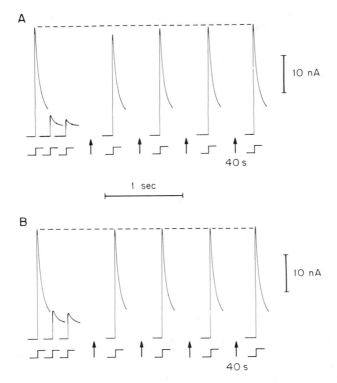

Figure 46. Rates of I_A decrease and recovery during the following repetitive command depolarizations. Current responses to only three of the first five depolarizing steps used are shown. The steps following arrows were given at 40-s intervals after the five depolarizing steps used for quantitation of differences. These five command depolarizations (2.2 s) to 0 mV occurred with a cycle time of 4.0 s. Each 2.2-s step was followed by a second command (800 ms) to +10 mV. (A) Command depolarizations paired with light. A light step (2.0 s) was presented 150 ms after the onset of each command depolarization. Light intensity: $10^{3.5}$ ergs/cm^2 s. (B) Command depolarizations alone. Arrows indicate 40-s intervals. Lower rectangular traces under A and B indicate onset of 60-mV command steps. Note that I_A (peak outward currents) decrease to lower values in (A) than in (B), and peak I_A takes minutes to return to the original values in (A) but not in (B). The first three currents included in (A) and (B) are the first, second, and fifth currents elicited by the five successive command steps. (From Alkon et al., 1982)

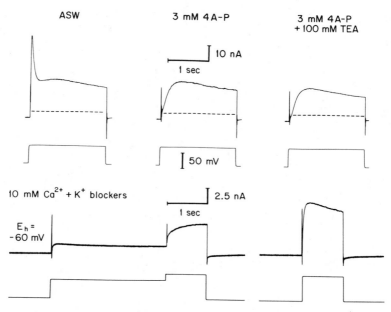

Figure 47. Voltage-dependent outward currents across the membrane of the isolated type B cell soma are shown. (Upper records) The bathing solutions are, from left to right, ASW, 3 mM 4-aminopyridine (4-AP) added to ASW, and 4-AP and 100 mM tetraethylammonium (TEA) added to ASW. Note that addition of 4-AP and TEA removes only a small portion of the late outward current elicited by command to 0 mV from a holding potential of -60 mV. (Lower records) Late outward current with and without preceding depolaring step to -20 mV. The dashed lines indicate that level of the nonvoltage-dependent or leak current. (From Alkon et al., 1984)

Repeated measurements of these currents for type B cells from conditioned and control animal. (Figure 51) reveals unequivocal conditioning-specific differences. The two major K^+ currents, I_A and $I_{Ca^{2+}-K^+}$, are substantially reduced (30–40%) for conditioned, as compared to other, animals (Figures 52 and 53). Furthermore, the magnitude of the current reduction for $I_{Ca^{2+}-K^+}$ is significantly correlated with the magnitude of the learning effect. Thus, the conditioning-induced change of a single current in a single identified neuron is related to the conditioning-induced behavioral change of the living animal. This clear conditioning-induced reduction of current flow through particular membrane channels accounts for the increased resistance and thus the increased excitability of the type B cells. Conditioning-induced reduction of I_A and $I_{Ca^{2+}-K^+}$, similar to the increased resistance, was observed for type B somata

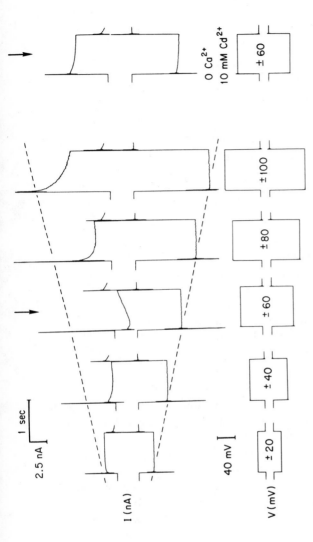

Figure 48. Records of voltage-dependent inward current present in a type B photoreceptor are shown under conditions of 100 mM external Ba²⁺ and K⁺ current blockers (10 mM 4-AP, 100 mM TEA). From a holding potential of −60 mV, successive command steps in multiples of 20 mV reveal activation of an inward current at potentials more positive than −40 mV. This current increases nonlinearly, reaching its peak value at 0 mV (left arrow), and diminishes thereafter. Note that at 0 mV (absolute), the inward current decreases substantially from its maximum amplitude. Such a decrease was eliminated when substitution of Ba²⁺ for Ca²⁺ in the ASW was preceded by a thorough washing with 0 Ca²⁺-ASW. From a holding potential of −60 mV, successive commands to potentials less than −60 mV elicit a nonvoltage-dependent leak current, indicated by the lower dashed lines. The upper dashed line is extrapolated from potentials less than or equal to −50 mV. Note that when 10 mM Cd²⁺ (which abolishes the Ca²⁺ current) is substituted for Ca²⁺ in the external bathing medium, the current elicited by the command to 0 mV (absolute) begins to approach the extrapolated leak level (indicated by the dashed line). (From Alkon et al., 1984)

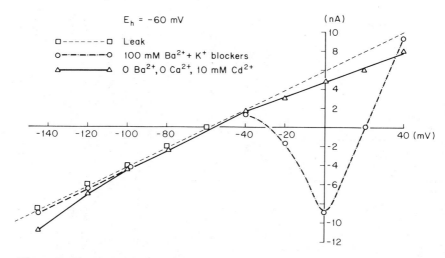

Figure 49. The steady-state current–voltage plot of the voltage-dependent inward current present in a type B photoreceptor is shown. Under conditions of high external Ba^{2+} (100 mM) and blockade of the voltage-dependent K^+ currents ($\bigcirc$), the current is inward over the range of $-40-+40$ mV (absolute), reaching its peak at ~ 0 mV. Removal of Ba^{2+} and Ca^{2+} from the bath, and addition of the calcium-channel blocker Cd (10 mM) after 5 min substantially reduced the inward current at all levels of membrane potential ($\triangle$). The residual deviations from the extrapolated leak current values ($\square$) may be caused by incomplete block of the inward current. (From Alkon et al., 1984)

isolated from all synaptic interaction and thus are intrinsic membrane changes. The current reduction persists, as does the learned association, for days after the conditioning experience.

It is important to be aware that the potassium currents, I_A and $I_{Ca^{2+}-K^+}$, are very small at the resting or unstimulated level of potential across the type B membrane. These currents only become significantly activated when the membrane potential becomes substantially more positive (e.g., by 30–60 mV) than the resting level. This means that a learning-induced reduction of I_A and $I_{Ca^{2+}-K^+}$ has no real effect on the type B cell membrane properties and its impulse activity at the resting potential level that occurs in the dark. But when the cell depolarizes in response to a light stimulus, I_A and $I_{Ca^{2+}-K^+}$ are activated. When the conditioned stimulus, light, is presented after conditioning, it evokes a new behavioral response. In the dark, as well as in response to sensory stimuli other than light, *Hermissenda* behavior is unchanged by prior training with paired light and rotation. The lack of a significant reduc-

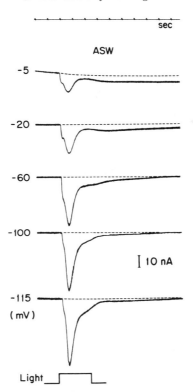

Figure 50. Light-induced inward Na$^+$ current. Light steps (10^4 ergs/cm^2 s), presented at 2-min intervals after 10 min of dark adaptation, elicited an inward current of constant amplitude at a holding potential of -60 mV. When the light occurred 5 s after the onset of a 20-s depolarizing command step, the peak amplitude of the large initial phase of the inward current (I_{Na^+}) became progressively smaller as the depolarizing command became more positive. A small, delayed (≥ 8 s from light onset) phase of the inward current, however, became larger with more depolarization. A delayed inward current ≤ 8 s from light onset), apparent at more negative holding potentials, may represent residual I_{Na^+}. The biphasic nature of the I_{Na^+} onset may be due to the presence of a second opposing light-induced current, $I_{Ca^{2+}-K^+}$. (From Alkon and Sakakibara, 1985)

tion of K$^+$ currents at the resting potential of conditioned type B cells (i.e., in the dark), is entirely consistent with a light stimulus presentation being *necessary* for behavioral expression of learning (as the light is necessary for expression of learning-induced differences of ionic currents). This lack of differences in the K$^+$ currents of resting-conditioned cells is also an electrophysiologic representation of an important aspect

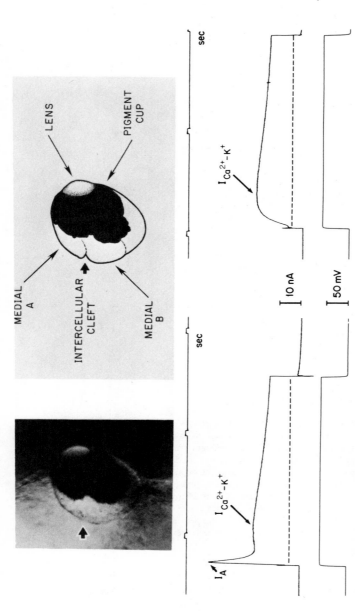

Figure 51. (A) Eye of the nudibranch mollusc *Hermissenda crassicornis*. The right (looking from a dorsal view) eye (∼ 60 μm in width) was isolated by axotomy from the circumesophageal nervous system and rotated 90° laterally and 90° anteriorly. The medial type A photoreceptor, located toward the lens (only faintly visible) from the longer medial type B photoreceptor occupying the caudal half of the intercellular cleft (to the left side of the black pigment cup) is separated by an transparent medial border to the left of the black pigment cup. The long axis of the eye is approximately 100 μm in length. Ionic currents across the membrane of the medial type B photoreceptor were measured for conditioned and control animals. (B) Voltage-dependent K⁺ currents measured in isolated type B soma I_A appears on the left as a fast early outward current ($V_H = -60$ mV) that is eliminated on the right in the presence of 3 mM 4-aminopyridine (4-AP) in the external bathing medium. I_C, measured 300 ms after the onset of the positive command step, is minimally affected by the 4-aminopyridine or as shown elsewhere (see Alkon et al., 1984) by 100 mM tetraethylammonium ion (TEA). (From Alkon et al., 1985)

82

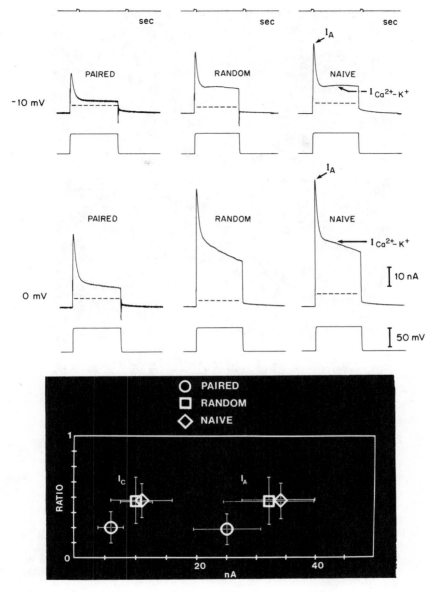

Figure 52. (A) Comparison of voltage-dependent K^+ currents measured in type B somata isolated from paired, random, and naive animals. The records were chosen to illustrate the reduction of I_A and $I_{Ca^{2+}-K^+}$ for paired, as compared to random and naive, animals. (B) Mean phototaxis suppression ratios in relation to ionic current magnitude. For individual animals of each group (paired, $\bigcirc$, random, $\square$; and naive, $\diamondsuit$) a suppression ratio (in the form $B/A + B$ where A = posttreatment latency and B = pretreatment latency) was obtained and the magnitude of $I_{Ca^{2+}-K^+}$ (on the left) and I_A was measured at -10 mV (absolute) across the isolated soma membrane of the medial type B cell. The values presented ($\pm$ SD) are the mean ionic currents ($I_{Ca^{2+}-K^+}$ and I_A) measured in relation to the mean suppression ratio for each group. The paired mean ratios and ionic currents are all clearly lower than for the random and naive groups. (From Alkon et al., 1985)

83

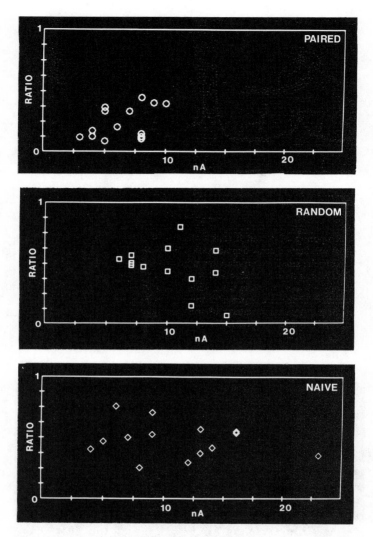

Figure 53. (A) *Paired*. Phototaxis suppression rate in relation to $I_{Ca^{2+}-K^+}$ magnitude for the paired group. For each animal a suppression ratio (in the form $A/A + B$) was obtained and the magnitude of $I_{Ca^{2+}-K^+}$ was measured at -10 mV (absolute) across the isolated soma membrane type B cell. The ratio is significantly correlated with the current magnitude. (B) *Random*. Phototaxis suppression ratio in relation to $I_{Ca^{2+}-K^+}$ magnitude for the random group. The ratio is not significantly correlated with the current magnitude. (C) *Naive*. Phototaxis suppression rate in relation to $I_{Ca^{2+}-K^+}$ magnitude for the naive group. The ratio is not significantly correlated with the current magnitude. (From Alkon et al., 1985)

of learned behavior – namely, the *recall* of a learned association (of which more will be said in the following chapter).

One way in which reduction of K^+ currents increased the excitability of type B cells from conditioned animals is, as discussed above, by increasing resistance to current flow. Another way becomes clear from an understanding of what happens to the potential of a cell during the flow of ions across its membrane. When positive ions such as potassium leave the inside of the cell, the inside of the cell becomes less positively charged (or more negatively charged). The more negative is the type B cell membrane potential (inside vs. outside), the farther away is its potential from that level critical for triggering impulses. For a more negatively charged type B cell, a greater positive shift of potential (i.e., depolarization) will be required to reach the threshold potential.

In summary, the flow of potassium ions to the outside of the cell makes the type B cell less excitable, whereas the flow of sodium or calcium ions to the inside of the cell makes it more excitable (i.e., brings the membrane potential closer to impulse threshold). The larger the potassium currents, I_A and $I_{Ca^{2+}-K^+}$, the less excitable is the type B cell. Conditioning results in a more excitable type B cell, therefore, by reducing I_A and $I_{Ca^{2+}-K^+}$.

It is perfectly conceivable that other changes of membrane currents should occur with learning. There is some evidence, for example, that with *Hermissenda* conditioning, I_A and $I_{Ca^{2+}-K^+}$ increase across the type A photoreceptor soma membrane. This makes the type A cell *less* excitable (vs. the type B cell, which has become *more* excitable). These are complementary membrane changes that have a similar effect on the passage of signals through a portion of the visual pathway (see above) involved in turning movement. Type B cells inhibit type A cells, which excite interneurons, which then excite motorneurons, which then cause turning. Less visually elicited turning will result from fewer type A impulses and more type B impulses in response to light.

Similarly, the K^+ currents need not be the only currents to change. An inward current such as $I_{Ca^{2+}}$ might change with learning and thereby affect a neuron's excitability. What the *Hermissenda* results teach us is that ionic currents do change and remain changed with learning. Ionic currents have long been known to change, of course, during a variety of electrophysiologic phenomena such as the impulse, a synaptic potential, or a sensory receptor (generator) potential. But these phenomena involve ionic current changes lasting anywhere from a few thousandths of a second to many seconds. The changes of ionic currents measured with the associative learning of *Hermissenda* last for days and perhaps much longer. This is an entirely new temporal domain for the physiology of ionic currents and it is this class of physiologic phenomena that may provide a special language of the nervous system beautifully designed for learning and memory.

Bibliography

Alkon, D. L. (1979). Voltage-dependent calcium and potassium ion conductances: a contingency mechanism for an associative learning model. *Science* 205:810–16.

   (1982–3). Regenerative changes of voltage-dependent Ca^{2+} and K^+ currents encode a learned stimulus association. *J. Physiol. Paris* 78:700–6.

Alkon, D. L., Lederhendler, I., and Shoukimas, J. J. (1982a). Primary changes of membrane currents during retention of associative learning. *Science* 215:693–5.

Alkon, D. L., Shoukimas, J. J., and Heldman, E. (1982b). Calcium-mediated decrease of a voltage-dependent potassium current. *Biophys. J.* 40:245–50.

Alkon, D. L., Farley, J., Sakakibara, M., and Hay, B. (1984). Voltage-dependent calcium and calcium-activated potassium currents of a molluscan photoreceptor. *Biophys. J.* 46:605–14.

Alkon, D. L., and Sakakibara, M. (1985). Calcium activates and inactivates a photoreceptor soma potassium current. *Biophys. J.* 48:983–95.

Alkon, D. L., Sakakibara, M., Forman, R., Harrigan, J., Lederhendler, I., and Farley, J. (1985). Reduction of two voltage-dependent K^+ currents mediates retention of a learned association. *Behav. Neural Biol.* 44:278–300.

Eccles, J. C. (1957). *The Physiology of Nerve Cells*. Johns Hopkins University Press, Baltimore.

Farley, J., and Alkon, D. L. (1982). Associative neural and behavioral change in *Hermissenda*: consequences of nervous system orientation for light- and pairing-specificity. *J. Neurophysiol.* 48:785–807.

Goh, Y., and Alkon, D. L. (1984). Sensory, interneuronal and motor interactions within the *Hermissenda* visual pathway. *J. Neurophysiol.* 52:156–69.

Goh, Y., Lederhendler, I., and Alkon, D. L. (1985). Input and output changes of an identified neural pathway are correlated with associative learning in *Hermissenda*. *J. Neurosci.* 5:536–43.

West, A., Barnes, E. S., and Alkon, D. L. (1982). Primary changes of voltage responses during retention of associative learning. *J. Neurophysiol.* 48:1243–55.

9

A regenerative origin for a new class of biophysical phenomena

The impulse, a fundamental element of the language of the nervous system, occurs within a fraction of a second. An enormously wide range of impulse frequencies provide biophysical expression for magnitude and duration of environmental events *in real time*. The impulse frequency changes as a function of the magnitude and duration of the stimulus. As the stimulus magnitude changes with time, so does impulse frequency. The stimulus input, of course, can be a sensory stimulus (such as light for the type B photoreceptor) or synaptic signals (from other neurons) that are elicited by an environmental stimulus.

Learning-induced transformations of membrane currents last at least many days and possibly weeks or longer. The relationship of these current transformations to the stimuli that produce them is quite different from the parallel dependence on time of impulse frequency and the stimuli that elicit the impulses. Stimuli may be repeated a number of times over the course of an hour's training period, but ionic current changes that occur as a result of learning persist long after the stimuli (the inputs) are gone. There is no longer a real-time parallel between input and neuronal response – there is no longer a direct or linear relationship between a stimulus and the biophysical effects of that stimulus (such as for impulses). On the contrary: A markedly nonlinear relationship – an exponential relationship – describes the dependence of learning-induced current changes on training stimuli. How does such nonlinearity arise? Can we reconstruct the sequence of cellular events that precede and inevitably lead to learning-induced biophysical phenomena that have a time course of days or longer rather than fractions of a second (as for the impulse)?

A conceptual framework for establishing such a sequence is provided by our present understanding of the impulse itself, because the origin of the impulse can also be traced to a sequence of cellular events that involve a markedly nonlinear relationship between a neuronal input and its biophysical effect.

Regenerative biophysical events during the impulse

As described earlier, when the voltage across a cell membrane reaches a critical threshold level, the impulse is triggered (Figure 42). This "triggering" is an "all-or-none" consequence of activating the flux of ions across the cell membrane. The flux of sodium ions across the membrane depends on the voltage in a very nonlinear manner. The flux rises as an exponential function of the voltage. What happens to make the impulse an "all-or-none" phenomenon is critically determined by the exponential increase of ion flux as the voltage across the cell membrane (inside vs. outside) becomes more positive. The impulse is also critically determined by a self-propelling or regenerative mechanism. As positive sodium ions rush across the membrane into the cell, the inside of the cell becomes more positively charged – the membrane potential more positive. With more positive membrane potentials, however, still more positive sodium ions rush into the neuron (because of the dependence of sodium flux on the membrane potential). With more sodium flux the membrane potential becomes still more positive triggering more sodium flux, and so on. This regenerative process of sodium flux and membrane potential enhancing each other is the essence of the explosive nature of the impulse and explains why the impulse is an inevitable consequence of reaching a threshold or triggering level of voltage across the cell membrane.

Over a time interval of some seconds, impulse frequency is a good expression of the real-time dependence of inputs to a neuron. This real-time dependence, however, is violated during the actual generation of the impulse itself. During the explosive generation of the impulse, there is a suspension of the direct one-to-one relation of biophysical events to the time dependence of the input that triggers the impulse. This suspension occurs within a time domain of thousandths of a second. A similar suspension occurs during learning but within an entirely new time domain – hours, days, and perhaps longer.

Regenerative biophysical events during learning an association of stimuli

To describe the regenerative production of learning-induced biophysical effects, we must trace, in a stepwise fashion, the biophysical and biochemical consequences, of exposing neural systems to particular stimulus relationships. The responses of individual neurons depend on the inputs (synaptic, and also for receptors, sensory) received from other neurons within the systems as well as their own membrane properties. The type B

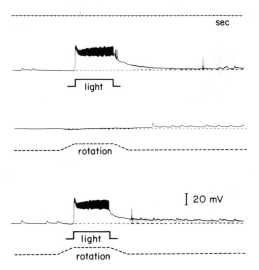

Figure 54. Responses of a type B photoreceptor to light and rotation. Rotation alone (middle panel) causes an intitial synaptic hyperpolarization, followed by prolonged depolarization, and increased frequency of excitatory postsynaptic potentials (EPSPs). Light alone (upper panel) causes a depolarizing response, followed by a slight increase of EPSP frequency (some increase is typical). The light and rotation stimuli when presented alone were of a somewhat longer duration than when presented together, that is, paired (lower panel). Note that the paired presentation is followed by a larger depolarization and a greater EPSP frequency than following either stimulus presented alone. (From Alkon, 1985)

photoreceptor, for example, depolarizes (i.e., its membrane potential becomes more positive) during a 30-s period of light (with the blue wavelength, 510 nm, which excites the visual pigment rhodopsin). The membrane potential, however, does not return to its previous negative level immediately after the light stimulus (of at least moderate intensity) goes off (Figures 54 and 81). Depending on the intensity and duration of the light, the type B membrane potential (inside vs. outside) remains more positive than its original level for one or more minutes after the light.

In response to rotation, the type B photoreceptor hyperpolarizes, that is, its membrane potential becomes more negative. The type B cell depolarizes immediately following, and for many seconds after, the rotation stops (Figure 54). This response of the type B cell to rotation is due to a number of synaptic inputs (Figure 55) from the vestibular or statocyst pathway (which is directly sensitive to rotation). Hair cells, for

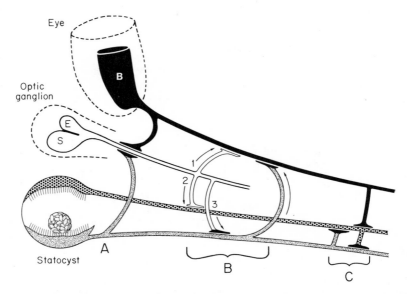

Figure 55. Intersensory integration by the *Hermissenda* nervous system. (A) Convergence of synaptic inhibition from type B and caudal hair cells on the E cell. (B) Positive synaptic feedback onto a type B photoreceptor. 1, direct synaptic excitation, 2, indirect excitation: E–S excites the cephalic hair cell that inhibits the caudal hair cell and thus disinhibits the type B cell; 3, indirect excitation: E inhibits the caudal hair cell and thus disinhibits the type B cell. 4, indirect excitation: the B cell inhibits the C cell, and thus disinhibits the E cell; C cell effects are not illustrated. (C) Intra- and intersensory inhibition. Cephalic and caudal hair cells are mutually inhibitory. The type B cell inhibits mainly the cephalic hair cell. All filled endings indicate inhibitory synapses; open endings indicate excitatory synapses. (From Tabata and Alkon, 1982)

instance, excited by rotation, depolarize, have more impulses, and thus release more inhibitory transmitter onto the type B cell – causing hyper-polarization. Hair cells after rotation actually hyperpolarize (i.e., the voltage from the inside to the outside of the membrane becomes more negative), have fewer than their resting number of impulses, and thus release less inhibitory transmitter onto the type B cell than *before* the rotation stimulus. After rotation, the type B cell is therefore released from the inhibition it normally receives due to the steady low level of impulse activity of unstimulated hair cells. Released from inhibitory, hyperpolarizing input after rotation, the type B cell depolarizes.

The type B cell also depolarizes following rotation for another reason: more excitatory synaptic transmitter is released from an optic ganglion cell. The E optic ganglion cell, like the type B photoreceptor, is also released from steady inhibitory input from the hair cells when the rotation stops (Figure 55). The E optic ganglion cell, when released from hair cell inhibition, releases more excitating transmitter onto the type B photoreceptor.

To recapitulate, the type B cell itself is not sensitive to rotation (as it is to light), but undergoes synaptic hyperpolarization *during* rotation and synaptic depolarization *following* rotation as a consequence of the stimulation of the vestibular pathway and as a consequence of the synaptic interactions between the visual and vestibular pathways.

When the light and rotation occur together, the type B cell responds differently than it does to either stimulus presented alone. Light paired with rotation is followed by larger and more prolonged depolarization (Figures 54 and 56). Depolarization of the type B cell *during* light alone is not different from depolarization during light paired with rotation. But the depolarization of the type B cell after the light is enhanced for the paired stimuli. If rotation precedes the light, this enhancement of type B depolarization (following the light) does not occur. Similarly, if the onset of rotation follows the onset of light by too great an interval, enhancement of type B depolarization does not occur. Enhanced type B depolarization then *only* occurs when the light and rotation are associated within a specific temporal interval. *Hermissenda* learn the association between light and rotation also only for the same specific temporal interval and order. Type B depolarization is a cellular expression of the behaviorally manifest *temporal* specificity of the *Hermissenda* associative learning. Type B depolarization is also cellular expression of behaviorally manifest *stimulus* specificity of the learning. When light is paired, for example, with rotation and the animal's head is confined so that it points toward the center of rotation, the animal learns to move less readily toward the light (as described above). When light is paired, however, with rotation and the animal's head is confined so that it points away from the center of rotation, the animal actually learns to move *more* readily toward the light. Amazingly, the membrane potential of the type B cell faithfully parallels these stimulus-specific differences in learned behavior. With the head-toward-the-center orientation, the type B cell responds with more depolarization, as already described, to light paired with rotation (i.e., depolarization after the stimuli is enhanced). With the head-away-from-the-center orientation, however, type B depolarization following light is *reduced* by pairing with light.

These features of stimulus and temporal specificity for the type B cell responses, as well as for the behavioral responses to light associated with rotation, are predicted by the known organization of the synaptic rela-

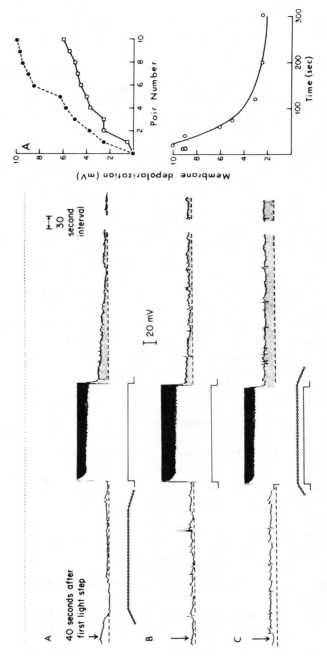

Figure 56. (Left) Intracellular voltage recordings of *Hermissenda* neurons during and after light and rotation stimuli: (Left panel) Responses of a type B photoreceptor to the second of two succeeding 30-s light steps (with a 90-s interval intervening). The cell's initial resting potential, preceding the first of the two light steps in (A), (B), and (C), is indicated by the dashed lines. Depolarization above the resting level after the second of the two light steps is indicated by the shaded areas. (A) Light steps (~ 10^4 ergs/cm^2 s) alternating with rotation (caudal orientation) generating ~ 1.0 g. The end of the rotation stimulus preceded each light step by 10 s. (B) Light steps paired with rotation. (C) Light steps alone. By 60 s after the second light steps, paired stimuli cause the greatest depolarization and unpaired stimuli the least. The minimal depolarization was, in part, attributable to the hyperpolarizing effect of rotation. Depolarization after the second presentation of paired stimuli was greater than that after the first. (From Alkon, 1980) (Right) Increase of type B membrane depolarization with repetition of the stimulus pairs. Membrane potential was measured instantaneously 20 s (●) and 60 s (□) after successive presentations of light steps paired with rotation. (B) Decrease of type B membrane depolarization after repeated presentations of stimulus pairs as described in (A). (From Alkon, 1980)

tions between neurons within and between the visual and statocyst pathways. The genetically determined neural system (Figure 55) defines the potential for stimulus and temporally specific cellular and behavioral responses to the association of light and rotation. The hair cells in the lower half of the statocyst, which are excited by rotation with head–center orientation), inhibit the type B photoreceptor and the E optic ganglion cell (Figure 56) explaining the synaptic effect of rotation on the type B cell described above. The hair cells in the upper half of the statocyst, which are excited by rotation with head-away-from-center orientation, do not inhibit either the type B or optic ganglion cells. These differences between synaptic input from hair cells located on opposite poles of the statocyst are critical for producing different synaptic responses of the type B cell to rotation and to light paired with rotation.

The stimulus and temporal specificity, however, do not simply depend on the organization of visual–statocyst synaptic interactions. The membrane properties of the type B cell itself make a significant contribution to this specificity. The prolonged depolarization of the type B cell after a light step is accompanied by a substantial increase in the "excitability" of the type B cell. Any current flowing across the type B membrane immediately *after* the light will cause a greater voltage change than usual because there is a greater resistance to current flow in general (Figure 57). By contrast, *during* the light there is a lower than normal resistance to current flow across the type B membrane – the type B cell is less "excitable" in response to other stimuli during a period of stimulation with its preferred stimulus, light, and more "excitable" following that period.

During the light step there is a net increase in the total number of open ion channels; *after* the light there is a net decrease in this number, even when compared to the cell before stimulation with light. A net increase of open channels means a lower than normal resistance to current flow across the type B membrane and a smaller voltage change; a net decrease means a greater than normal resistance to current flow and larger voltage change. What are the implications of high- and low-input resistance during different portions of the type B cell's response to light? In general, synaptic effects during a period of high-input resistance will be amplified (Figure 57), whereas synaptic effects during a period of low-input resistance will be reduced. This is because the synaptic effect produces a current across a particular locus of the type B membrane, and, when the input resistance of the whole cell is low, a smaller voltage

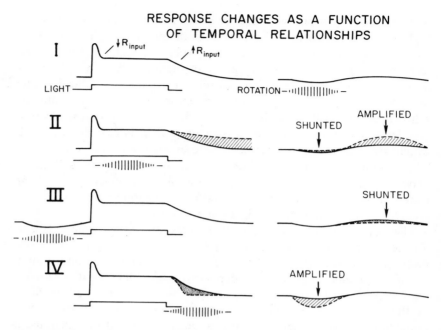

Figure 57. Effects of the temporal relationship of light and rotation stimuli on type B photoreceptor responses. Schematic responses to light stimuli (bars) are on the left; responses to rotation stimuli (sequence of vertical lines) are on the right. During light input, resistance (R_{input}) of the type B cell is low, thus shunting synaptic current due to activation of statocyst synaptic effects by rotation. Immediately after light, input resistance is two to three times higher than that prior to light, thereby amplifying rotation-induced synaptic input. I. Explicitly unpaired light and rotation stimuli. II. Paired stimuli. III. Light immediately preceded by rotation. IV. Rotation immediately following light. (From Alkon, 1986)

change will result from the flow of that synaptic current. The opposite occurs when input resistance is high.

During rotation (head–center orientation), the type B cell undergoes synaptically induced hyperpolarization. After rotation the type B cell undergoes synaptic depolarization. When light is paired with rotation, the synaptic hyperpolarization occurs during a period of low resistance across the type B membrane. The rotation-induced hyperpolarization is actually eliminated (or "shunted") when it occurs during this period of low type B resistance. However, when light is paired with rotation, the synaptic depolarization (following rotation) occurs during a period when type B input resistance is high. The rotation-induced synaptic depolarization is, therefore, enhanced after stimulus pairing by the membrane properties of the type B cell (Figure 57). When rotation occurs, in relation to when light occurs, will determine which part of the rotation's synaptic effects will be enhanced or reduced by the type B cell's membrane properties. Light paired with rotation eliminates synaptic hyperpolarization and enhances synaptic depolarization. Light followed by rotation enhances synaptic hyperpolarization with little effect on synaptic depolarization. Light preceded by rotation eliminates synaptic depolarization with little effect on synaptic hyperpolarization.

Interestingly, it is not only that the type B membrane properties during and after a light affect the rotation-induced synaptic effects, it is also vice versa: The rotation-induced synaptic effects influence the type B cell's membrane response during and after a light. Synaptic excitation during the period immediately following a paired presentation of light and rotation will cause the voltage across the type B membrane (inside vs. outside) to become somewhat more positive (by 5–10 mV). But the ionic currents responsible for the prolonged depolarization of the type B cell after light are further activated by this positive increment of voltage. Further activation of these voltage-dependent currents (including a calcium current and potassium currents) and a consequent further in-activation of the potassium currents leads to a further increase of input resistance (by a process described on pages 101 and 102. A further increase of input resistance will result in a greater voltage change due to synaptic excitation. Larger amplitudes of synaptic excitation will activate the membrane currents still further, and so on.

So with a single pairing of light and rotation, a stimulus- (and temporally) specific response results from the mutually enhancing inter-action of the synaptic organization with type B membrane properties. And in this mutually enhancing interaction, the regenerative or "explo-sive" nature of the type B membrane changes with learning is first suggested (Figure 58). With one pairing, the regenerative aspect of the type B cell depolarization is obviously limited. With repeated pairings (Figure 56), however, just as occurs during a normal session for training intact *Hermissenda*, the type B cell depolarization becomes progressively larger (i.e., it accumulates). And now, with repeated pairings, there is an additional element in the regenerative process just outlined. The type B depolarization that follows each pairing of light and rotation never entirely disappears before the next pairing (say after 90 s) occurs (Figure 56). So for the next response the type B cell is already slightly de-polarized and thus those currents activated by depolarization are further enhanced causing still more depolarization, further increased input resis-

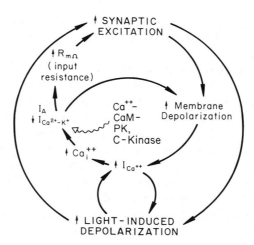

Figure 58. Regenerative synaptic- and light-induced excitation of the type B photoreceptor. Light-induced depolarization facilitates synaptic excitation and vice versa in response to temporally associated light and rotation. Analyzed in biophysical terms, synaptic depolarization causes transient activation and then prolonged inactivation of I_A and $I_{Ca^{2+}-K^+}$ and enhancement of a voltage-dependent Ca^{2+} current. Increased intracellular Ca^{2+} causes further inactivation of I_A and $I_{Ca^{2+}-K^+}$ and thus a further increase of effective input resistance. These in turn cause more membrane depolarization. Inactivation of I_A and $I_{Ca^{2+}-1K^+}$ by elevated intracellular Ca^{2+} may occur via the increased activity of Ca^{2+}–calmodulin-dependent protein kinase (Ca^{2+}–CaM-PK) and C-kinase. (From Alkon, 1984)

tance, etc. (Figure 57). After 10 light–rotation pairings the positive shift of the type B membrane potential can be quite substantial (e.g., 10–15 mV above the cell's resting potential level). Cumulative depolarization will persist for many minutes after 10 light–rotation pairings (Figure 56). It may persist for 1–2 h after 30 such pairings. Thus, the synaptic responses, activation of type B membrane currents, and the repetition effects of stimulus pairings enhance each other in a regenerative fashion to produce progressive membrane depolarization. The outcome of these regenerative interactions does not result in a very rapid or "explosive" change of membrane potential (as occurs during the impulse) – a nonlinear or exponentially increasing rate of potential change. Rather, the nonlinearity is manifest in the time for *recovery* of the depolarizing shift of membrane potential. The number of stimulus pairings is related in a nonlinear manner to the persistence of type B depolarization as well as to the persistence of increased type B input resistance.

The process of cellular transformations during learning in one temporal domain (i.e., hours, days, or years) might be usefully likened to geological transformations in a vastly longer temporal domain (i.e., thousands and millions of years). During acquisition, depolarization accumulates – it repeatedly increases, then decreases but never to its former level. With each repetition the membrane potential starts at a somewhat more positive level, until it reaches a level from which it takes many minutes or hours to recover. Similarly, during the formation of glaciers, snow accumulates each winter and melts each summer, but never melting enough to dissipate entirely the previous winter's accumulation. With each annual cycle, the forming glacier starts the winter with a somewhat greater mass of ice and snow, until it reaches a level from which it takes eons before being greatly reduced by a drastic climate change.

Now, both accumulated depolarization and glacier formation are reversible, but they may produce consequences that are much more irreversible. Glaciers, with their accumulated mass, can transform geological landscapes by mechanical force as well as by water released during melting. Similarly, prolonged depolarization together with accompanying Ca_i^{2+} elevation results in further profound changes of membrane and biochemical cellular properties – changes, which far outlast the depolarization.

Bibliography

Alkon, D. L. (1979). Voltage-dependent calcium and potassium ion conductances: a contingency mechanism for an associative learning model. *Science* 205:810–16.

(1980a). Cellular analysis of a gastropod (*Hermissenda crassicornis*) model of associative learning. *Biol. Bull* 159:505–60.

(1980b). Membrane depolarization accumulates during acquisition of an associative behavioral change. *Science* 210:1375–6.

(1982–3). Regenerative changes of voltage-dependent Ca^{2+} and K^+ currents encode a learned stimulus association. *J. Physiol. Paris* 78:700–6.

(1984). Calcium-mediated reduction of ionic currents: a biophysical memory trace. *Science* 226:1037–45.

(1985). Changes of membrane currents and calcium-dependent phosphorylation during associative learning. In *Neural Mechanisms of Conditioning*, ed. by D. L. Alkon and C. D. Woody, pp. 3–18. Plenum, New York.

Alkon, D. L., and Grossman, Y. (1978). Long-lasting depolarization and hyperpolarization in eye of *Hermissenda*. *J. Neurophysiol.* 41:1328–42.

Grossman, Y., Schmidt, J. A., and Alkon, D. L. (1981). Calcium-dependent potassium conductance in the photoresponse of a nudibranch mollusk. *Comp. Biochem. Physiol* 68:487–94.

Schmidt, R. F. (1978). *Fundamentals of Neurophysiology*. Springer-Verlag, New York.

Tabata, M., and Alkon, D. L. (1982). Positive synaptic feedback in the visual system of the nudibranch mollusc *Hermissenda crassicornis*. *J. Neurophysiol* 48:174–91.

10

Translation of psychological into biophysical phenomena

The limiting conditions that define associative learning at the behavioral level must also have meaning at the cellular level. There should be a translation of psychological features into properties of cellular responses. The origin of stimulus and temporal specificity in the regenerative depolarization of the type B cell already has provided evidence of such psychological–cellular translation. An animal's ability to learn the association between light and rotation depends on the light and rotation occurring within a certain time interval. If rotation precedes light or follows light by too long an interval or occurs at random in relation to the light, the animal fails to learn the association. The enhanced depolarization of the type B cell is similarly dependent on the temporal association of the light and rotation. The behavioral and neuronal responses to other stimuli not presented are not modified by paired light and rotation stimuli, that is, the learning behavior and the cellular responses are stimulus-specific. Stimulus specificity of cellular and behavioral responses was also apparent for rotation in opposite orientations. Type B depolarization is enhanced and the velocity of movement toward a light decreases when rotation in the head-toward-the-center orientation is paired with light. Type B depolarization is reduced and light-directed movement increases for rotation in the opposite orientation. Thus, via the integrated responses of the visual–statocyst neural systems, temporal and stimulus specificity of the learning behavior (i.e., at the psychological level) is translated into depolarization of the type B cell (i.e., at the cellular level). Similarly, at the behavioral level, the animal's learning of the association increases as a function of practice, that is, it shows acquisition. The learning improves with repetition of the light–rotation pairings. At the cellular level, type B depolarization (together with increased resistance) progressively accumulates as the pairings are repeated. Acquisition expressed in psychological terms is translated into cellular terms: cumulative depolarization.

98

What is the cellular expression of retention of the learned association (i.e., memory, assessed days after the training)? The membrane potential of the type B cell is no longer more positive than its previous resting level – it is no longer depolarized. But its resistance to current flow remains elevated – particularly current flow that makes the type B membrane potential more positive. The type B cell, on days after acquisition of the learned association, is more excitable, that is, current flow across its membrane will cause a greater voltage change than previously (and as compared to voltage changes of cells from control animals). Light elicits a larger depolarization of the type B cell, and, via the longer type B cell's inhibitory effect on certain visual interneurons and motorneurons, and excitatory effect on other interneurons and motorneurons decreased movement toward light. Retention of the learned association, at the behavioral level, is translated into increased type B resistance and excitability at the cellular level. Recall (remembering the association) at the behavioral level, is translated at the cellular level into a larger type B depolarization in response to light due to the increased excitability of the type B cell. As outlined in an earlier chapter, increased input resistance (and thus excitability) arises from reduced K^+ currents in type B cells from conditioned animals.

Since these K^+ currents are only significantly activated when the membrane potential of the type B cell is much more positive than its resting potential, differences in conditioned type B responses will only become manifest with a light stimulus at sufficient intensity. Thus, recall at the memory at the behavioral level is translated into voltage-dependent activation of the K^+ currents at the cellular level.

Thus the translation of psychological into cellular retention can be extended to the level of ionic channels, and it is on this level that we can trace a sequence of biophysical events that ultimately are responsible for the origin of this memory trace in *Hermissenda*.

Bibliography

Alkon, D. L. (1980). Membrane depolarization accumulates during acquisition of an associative behavioral change. *Science* 210:1375–6.

Alkon, D. L., Lederhendler, I., and Shoukimas, J. J. (1982). Primary changes of membrane currents during retention of associative learning. *Science* 215:693–5.

Alkon, D. L., Sakakibara, M., Forman, R., Harrigan, J., Lederhendler, I., and Farley, J. (1985). Reduction of two voltage-dependent K^+ currents mediates retention of a learned association. *Behav. Neural Biol.* 44:278–300.

Farley, J., and Alkon, D. L. (1980). Neural organization predicts stimulus specificity for a retained associative behavioral change. *Science* 210:1373–5.

Goh, Y., Lederhendler, I., and Alkon, D. L. (1985). Input and output changes of an identified neural pathway are correlated with associative learning in *Hermissenda*. *J. Neurosci.* 5:536–43.

Lederhendler, I., and Alkon, D. L. (1986a). Temporal specificity of the CS–UCS interval for *Hermissenda* Pavlovian conditioning. *Soc. Neurosci. Abstr.* 12:40.

Lederhendler, I. I., and Alkon, D. L. (1986b). Classical conditioning of *Hermissenda*: origin of a new response. *J. Neurosci.* 6:1325–31.

Lederhendler, I., and Alkon, D. L. (1986). Implicating causal relationships between cellular function and learning behavior. *Behav. Neurosci.* 6:833–838.

West, A., Barnes, E. S., and Alkon, D. L. (1982). Primary changes of voltage responses during retention of associative learning. *J. Neurophysiol.* 48:1243–55.

11
A biophysical sequence

All of those behavioral features which translate into depolarization of the type B cell, are, in turn, translated into the level of an intracellular "messenger" – calcium. During a light step, the level of intracellular calcium rises and stays elevated for many seconds (Figure 59). When the light is paired with rotation, calcium elevation is enhanced and prolonged. The concentration of calcium inside the type B cell increases for two reasons: Light releases calcium from storage sites within the cell; and depolarization activates a voltage-dependent flux of calcium ions across the type B membrane. It is the second mechanism that is most responsible for increased concentration of calcium within the cell. The greater, longer, and more frequent is the depolarization of the type B cell, the more intracellular calcium is elevated. This relationship of intracellular calcium to membrane potential is aided by an important feature of the flux of calcium across the membrane: it is sustained. The calcium flux or current does not inactivate. As long as the membrane potential is maintained at a certain level, the movement of calcium ions across the type B membrane continues. This enables the intracellular calcium concentration to more faithfully reflect the degree to which the type B cell is depolarized.

Artificial elevation of intracellular calcium can be accomplished by injection of calcium through a microelectrode inserted inside the type B cell. A single injection of calcium through one microelectrode results in a prolonged (many minutes) reduction of K^+ currents measured with two other previously inserted microelectrodes. The same two K^+ currents, which stay reduced on days after conditioning, are also reduced by calcium injection. These K^+ currents, I_A and $I_{Ca^{2+}-K^+}$, are also reduced after the type B cell is exposed to light paired with a period (e.g., 30 s) of depolarization. Light paired with depolarization simulates the effect of the visual and statocyst neuronal systems' response to light paired with rotation. Injection into the cell of a substance that binds or chemically

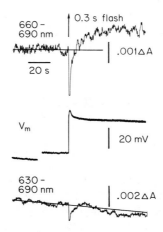

Figure 59. Absorbance changes for a 660–690 nm wavelength difference (top record) monitors the elevation of intracellular Ca^{2+} during light-elicited depolarization (middle record). Absorbance changes for a 630–690 nm wavelength difference (bottom record) measures pH or hydrogen ion concentration, which is unaffected following a 0.3-s light flash occurring at arrow. (From Connor and Alkon, 1984)

combines with calcium, called a calcium chelator, prevents elevation of intracellular calcium and also prevents the effect of light paired with depolarization on the type B cell K^+ currents. Prevention of calcium elevation prevents reduction of I_A and $I_{Ca^{2+}-K^+}$. Similarly, prevention or reduction of the flux of calcium ions across the type B membrane decreases or eliminates the K^+ current reduction following light paired with depolarization.

These observations, taken together, indicate that the level of intracellular calcium provides a link between the process of acquisition of the learning and the process of retention. Stimulus- and temporally specific depolarization of the type B cell accumulates during acquisition – during repeated presentations of light and rotation. Accompanying this depolarization is intracellular calcium elevation, which is greatest during each stimulus pairing. Elevated intracellular calcium causes prolonged reduction of K^+ currents and results in an increased resistance and "excitability" of the type B cell. On days after acquisition the depolarization of the type B cell and elevation of intracellular calcium are gone, but the increased resistance, the reduction of K^+ currents, persist. Elevated calcium has left a trace of its occurrence – a record expressed by the persistent reduction of I_A and $I_{Ca^{2+}-K^+}$.

Bibliography

Alkon, D. L., Shoukimas, J., and Heldman, E. (1982). Calcium-mediated decrease of a voltage-dependent potassium current. *Biophys. J.* 40:245–50.

Alkon, D. L., Farley, J., Sakakibara, M., and Hay, B. (1984). Voltage-dependent calcium and calcium-activated potassium currents of a molluscan photoreceptor. *Biophys. J.* 46:605–14.

Alkon, D. L., and Sakakibara, M. (1985). Calcium activates and inactivates a photoreceptor soma potassium current. *Biophys. J.* 48:983–95.

Connor, J. A., and Alkon, D. L. (1984). Light- and voltage-dependent increases of calcium ion concentration in molluscan photoreceptors. *J. Neurophysiol.* 51:745–52.

Sakakibara, M., Alkon, D. L., DeLorenzo, R., Goldenring, J. R., Neary, J. T., and Heldman, E. (1986a). Modulation of calcium-mediated inactivation of ionic currents by Ca^{2+}/calmodulin-dependent protein kinase II. *Biophys. J.* 50:319–27.

Sakakibara, M., Alkon, D. L., Neary, J. T., Heldman, E., and Gould R. (1986b). Inositol triphosphate regulation of photoreceptor membrane currents. *Biophys. J.* 50:797–803.

12
Molecular regulation of membrane channels during learning

The translation of behavioral aspects of learning into cellular physiology occurs in several temporal domains. Pairing specificity was manifest in the visual–vestibular network response over a period of seconds. Improvement of the learning, as translated into depolarization of the type B cell, occurred over a period of many minutes. Retention of the learning, as encoded by the persistent reduction of K^+ channels, lasted at least for many days. The transition of cellular phenomena in one temporal domain (e.g., for seconds and minutes) into another (e.g., for days or weeks) would seem intuitively to require some fundamental cellular transformations – transformations of the cell's structure or the basic biochemical machinery necessary for the ongoing renewal of the cell's essential constituents. The intuition of biochemical mediation of transitions between temporal domains arises from what we already know about cellular functions. First, we know that ultimately all cellular physiology is based on the chemical interaction of molecules. A persistent biophysical change must have some biochemical expression. We also know that under certain circumstances, such as during the growth and development of an organism, cells undergo radical and permanent transformations that are, to a significant degree, programmed in the genetic material, that is, they are specified in the DNA. Inevitably, the programmed information within the DNA is "read out" and expressed by the molecular machinery of the cell (i.e., by a long series of biochemical reactions). We also know that many molecules and subcellular structures have only a transient existence: They are degraded and recycled and then resynthesized. If a learning-induced change lasts for many days, then it has to outlast many of the cellular constituents that are turning over. There are a number of ways, biochemically, that relative invulnerability to turnover (or the synthesis–degradation cycle) might be accomplished.

104

A biochemical reaction may simply produce a molecule that has a very long half-life (i.e., survival time before its degradation). These occur in nature, although not commonly. Another possibility concerns the cell's biochemical machinery for manufacturing its constituents. The design for that manufacture, residing within the DNA (in the nucleus), might be altered during learning so that subsequent synthesis of specific proteins is different in certain neurons. Alternatively, transmission (by messenger RNA) of the design to the RNA (in the cytoplasm) where the proteins are assembled might be affected during learning. Or the assembly process itself could be a site for learning-induced change. Furthermore, a biochemical way in which degradation invulnerability could be conferred might depend upon ongoing signaling between the cytoplasm (where proteins are synthesized) and the nucleus (where protein design is specified). When the state of part of a protein changes with learning, the new protein state could signal the nucleus to produce more of these proteins in the same state. This would be, in effect, a feedback cycle – similar to the regenerative processes that, as we discussed, underlie the impulse as well as the changes of the type B biophysical properties during the acquisition of a learned association between the light and rotation stimuli. Such a feedback cycle need not be thought of as possible only for nuclear–cytoplasmic interaction. It might occur within only one cellular compartment (e.g., the cytoplasm). An altered protein state could have the effect of causing still more of the same altered protein states. Once a certain number of these altered protein states arose, a threshold might be reached so that the process would be, to some degree, self-perpetuating. A good example of this might be provided by activation of an enzyme that facilitates molecular interaction, which in turn activates this enzyme still further or in a more persistent manner.

Since injection of calcium into the type B cell caused reduction of K^+ currents similar to that observed with conditioning, biochemical steps sensitive to calcium elevation are a reasonable starting point for a search into molecular mechanisms of ionic channel regulation during learning.

Calcium is known in a variety of physiologic settings as a "second messenger." Signals arising external to a cell cause elevation of calcium levels inside the cell. Altered intracellular calcium in turn carries the "message" to other intracellular constituents by stimulating particular biochemical pathways and thereby producing physiologic effects such as secretion of hormones, aggregation of blood-clotting elements called platelets, or muscular contraction. There are several major biochemical routes that calcium can stimulate to cause its effects. Expression of these effects often requires the addition of phosphate groups onto proteins. For example, calcium can work together with another molecule, called calmodulin, to stimulate an enzyme, called a kinase, which in turn facilitates phosphorylation. Calcium–calmodulin-dependent phosphorylation of structural proteins, called microtubule-associated proteins, is one example. Calcium can also work together with in the presence of other molecules called phospholipids, diacylglycerol, a "fatty" substance, to activate an entirely different kinase called "C-kinase." C-kinase-facilitated phosphorylation regulates secretion of a hormone (aldosterone) from cells in the adrenal gland, smooth muscle contraction, and platelet aggregation. Calcium can also stimulate still another type of enzyme called adenylate cyclase. Stimulation of adenylate cyclase raises the intracellular level of another "second messenger," called cyclic-AMP (short for cyclic-adenosine monophosphate), which also activates a kinase and thereby stimulates phosphorylation. Calcium does not only trigger enzymatic pathways responsible for the addition of phosphate groups, it also can facilitate breaking or cleavage of phosphate bonds to proteins by stimulating enzymes called phosphatases and proteases.

Several lines of experimental evidence implicate calcium–calmodulin-dependent and calcium–lipid-dependent phosphorylation in the generation of learning-induced reduction of K^+ currents. Injection of a molecule (inositol-trisphosphate) responsible for releasing calcium from intracellular storage depots into type B cells is, in fact, followed by reduction of the same two K^+ currents, I_A and $I_{Ca^{2+}-K^+}$, which remain reduced after classical conditioning (Figure 60). The specificity of the inositol-stimulated reduction of currents is remarkably similar to, if not identical with, that of the current reduction of classical conditioning. Whereas the I_A and $I_{Ca^{2+}-K^+}$ are reduced, the light-induced I_{Na^+} is unaffected by either the inositol injection or the learning. Injection of a closely related molecule, inositol-monophosphate, which is without the biological capability of releasing intracellular calcium, has no effect on the type B membrane currents.

Other experimental manipulations provide further support for the interpretation that inositol-trisphosphate acts through the calcium–calmodulin-dependent phosphorylation pathway. Injection of a calcium–calmodulin-dependent kinase into the type B cell is at first without effect. If, however, the type B cell is then exposed to a period (e.g., 20 s) of a positive change of membrane potential paired with light,

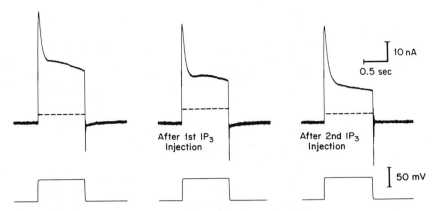

Figure 60. Effect of IP_3 injection on voltage-dependent outward currents across the membrane of the isolated type B photoreceptor cell soma. The currents were elicited by commands to -5 mV (absolute) from a holding potential of -60 mV. I_A and $I_{Ca^{2+}-K^+}$ were markedly reduced after IP_3 injection (-2.0 nA for 2 min) and further reduction occurred after a second injection. This reduction persisted for the duration of the recording. The dashed line indicates the level of the nonvoltage-dependent or "leak" current. (From Sakakibara et al., 1986b)

again the same two K^+ currents, which are reduced with conditioning (I_A and $I_{Ca^{2+}-K^+}$), are decreased and remain decreased at least for a few hours later (Figure 61). What this means is that the enzyme injection is without effect until a rise of intracellular calcium occurs within the type B cell. The K^+ current reduction that ordinarily persists for 2–3 min after this rise of intracellular Ca^{2+} is greatly prolonged after the kinase injection. The elevation of intracellular calcium necessary to activate the enzyme results from flux across the membrane due to activation of the voltage-dependent calcium current by the positive shift of membrane potential. A smaller calcium elevation also results when the light (paired with the shift of membrane potential) releases calcium from intracellular storage depots. The marked prolongation of K^+ current reduction does not occur after injection of the same enzyme, which has been inactivated by standing at room temperature. Similar to the learning-induced reduction of K^+ currents, that induced by calcium loading together with kinase injection is again not accompanied by a change of the light-induced inward Na^+ current, I_{Na^+}. If the type B cell is first bathed in an inhibitor (trifluoperazine) of calcium–calmodulin-dependent and C-kinase-dependent phosphorylation, the injection of Ca^{2+}–calmodulin-dependent kinase has no effect (Figure 62). Furthermore, the trifluoperazine

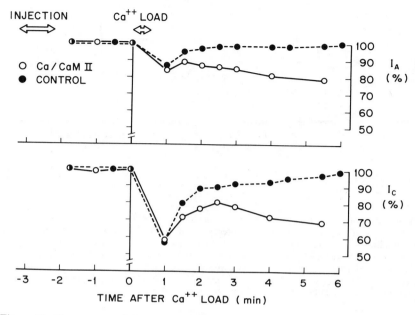

Figure 61. Time course of I_A (upper panel) and $I_{Ca^{2+}-K^+}$ (lower panel) reduction following a Ca^{2+} load. I_A and $I_{Ca^{2+}-K^+}$ (or I_C) were measured as the peak outward currents ~ 20 ms (I_A) and 300–400 ms (I_C) from the onset of a command depolarization to −5 mV (absolute) (holding potential = −60 mV). Ca^{2+} loads were given before (●) and after (○) iontophoretic injection of CaM kinase II. I_A and I_C amplitudes before the Ca^{2+} load are normalized at 100%. Note that CaM kinase II injection prevents recovery of I_A and I_C reduction after a Ca^{2+} load. (From Sakakibara et al., 1986a)

causes an increase of I_A and $I_{Ca^{2+}-K^+}$, suggesting that inhibition of enzyme activity already present in the cell regulates the currents. Thus, injection of the kinase (with a Ca^{2+} load) *simulated* learning just as injection of inositol-trisphosphate (without a Ca^{2+} load) did. Presumably, a Ca^{2+} load was not necessary for inositol-trisphosphate to cause K^+ current reduction, since inositol-trisphosphate releases calcium from intracellular stores.

Simulation of the biophysical effects of learning also follows activation of another kinase when coupled with a Ca^{2+} load. The activity of C-kinase, which is activated both by Ca^{2+} and diacylglycerol, is greatly enhanced after exposure to a particular tumor-producing agent called phorbol ester which acts like diacylglycerol. Bathing the type B cell in phorbol ester is sufficient to activate the C-kinase. Again, Ca^{2+} loading after the phorbol ester treatment is followed by persistent reduction of

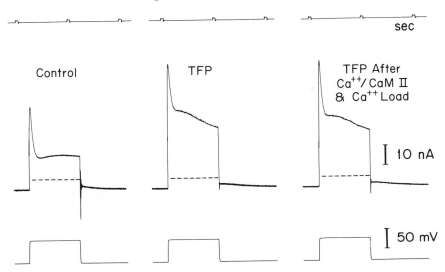

Figure 62. Trifluoperazine (TFP) perfusion increases amplitudes of outward K^+ currents (I_A and I_C). TFP also markedly reduces or eliminates the effect of CaM kinase II injection on I_A and I_C with Ca^{2+} loads. (From Sakakibara et al., 1986a)

the K^+ currents, I_A and $I_{Ca^{2+}-K^+}$ (Fig. 63), without significant change of the light-induced current, I_{Na^+}. The effects follow injection of C-kinase into the type B cell provided the C-kinase is first activated by phorbol ester treatment of the cell.

What may occur during the acquisition of classical conditioning, therefore, is activation of both the Ca^{2+}–calmodulin-dependent and Ca^{2+}–lipid-dependent kinases to affect phosphorylation of proteins that either regulate, or are a part of, K^+ channels within neuronal membranes (Fig. 58). In fact, these two kinases may act synergistically to produce effects greater and more prolonged than those that follow activation of either kinase alone. Furthermore, activation of each of these phosphorylation pathways may be due more exclusively to the conditioned stimulus, light, or the unconditioned stimulus, rotation. Light (the CS), for example, has been shown to increase intracellular levels of the lipid molecule, diacylglycerol, in other invertebrate photoreceptors and thereby could activate the C-kinase. It may be necessary that the CS first activates the C-kinase, which then is "primed" to be more sensitive to Ca^{2+} elevation due to CS–UCS pairing. A precedent for such synergistic interaction is amply provided in the endocrine system. Platelet aggregation is enhanced and prolonged to the greatest degree

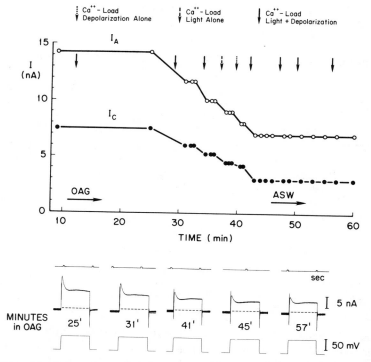

Figure 63. Effects of exposure to OAG (a C-kinase activator) on K⁺ currents across type B soma membrane. Successive calcium loads (indicated by arrows) cause progressive persistent reduction of K⁺ currents, I_A and $I_{Ca^{2+}-K^+}$ (I_C), after $\geq$ 30 minutes of OAG treatment. A graphic representation of the actual current values (after "leak" correction) is shown in the upper panel. The cell was penetrated and voltage-clamped with two microelectrodes 9 min after exposure to OAG was begun. Lower records (for the same cell) show actual currents elicited at the times shown and after Ca²⁺ loads indicated above. Note that Ca²⁺ loads were not followed by further persistent reduction of K⁺ currents after OAG was removed from the ASW (at 43 min) nor did the K⁺ currents recover their former amplitude in ASW. (From Alkon, et al., 1986)

only when both the Ca²⁺–calmodulin-dependent and Ca²⁺–lipid-dependent kinases are activated. Similarly, secretion of the hormone aldosterone by adrenal cells is only maximal when both kinases have been activated, and in this case also, there seems to be a requirement for a particular order and timing of activation of each of the two kinases.

Consistent with these biophysical results of kinase activation was the finding that calcium-dependent phosphorylation of a low molecular weight protein (i.e., 20,000 mol. wt.) changed in cells of eyes isolated

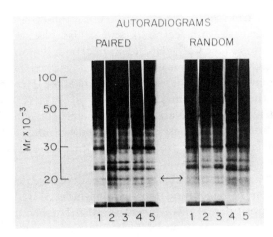

Figure 64. SDS electrophoresis gels comparing endogenous protein phosphorylation in the eyes of *Hermissenda* presented with paired or random light and rotation. Additional groups of eyes from *Hermissenda* presented light alternating with rotation or given no training (i.e. naive) showed phosphorylation profiles essentially identical with those for the randomized light and rotation group. There is a change in the level of incorporation of ^{32}P in a 20,000-mol. wt. phosphoprotein. Since the eye of *Hermissenda* consists of only five photoreceptors, a lens and a few pigment and epithelial cells, a biochemical change specific to associative learning has thus been localized to a few cells within a nervous system. (From Neary et al., 1981)

from conditioned *Hermissenda* but not from control animals (Figure 64). This conditioning-specific change of phosphorylation was measured with a technique, called microgel analysis, that exposes photographic plates to proteins that have incorporated radioactive phosphate into their structures. The proteins were separated on a thin proteinaceous slab, called a gel, by their different abilities to move within an electric field.

This technique, called gel electrophoresis, also revealed differences of phosphorylation when *Hermissenda* nervous systems were exposed to prolonged depolarization (as occurs to the type B cell during acquisition of a learned association). Prolonged depolarization is produced by bathing the nervous systems in sea water with a high concentration of K^+ ions. Depolarization for 10–30 min was followed by a reduction in phosphorylation of low molecular weight proteins (25,000 and 20,000 mol. wt.). This reduction lasted for 30–60 min *after* the depolarization conditions (i.e., high external K^+) have been removed. The reduction of 25,000 and 20,000 mol. wt. phosphorylation was prevented by the inhibition of Ca^{2+}–calmodulin-dependent and Ca^{2+}–lipid-dependent phosphorylation with trifluoperazine. Thus, this inhibitor blocked the action of Ca^{2+}–calmodulin-dependent kinase injection to reduce the K^+ currents (as occurred with learning), and it also blocked the action of depolarization to reduce Ca^{2+}-dependent phosphorylation of specific low molecular weight proteins measured with gel electrophoresis. A close parallel was demonstrated, therefore, between (1) the effect of associative learning on the K^+ currents (I_A and $I_{Ca^{2+}-K^+}$); (2) the effect of Ca^{2+}-dependent kinase activation on the same currents; and (3) the effect of Ca^{2+}-dependent kinase activation and learning on phosphorylation of low molecular weight proteins within the cytosolic and membrane compartments of the neuron.

It should be emphasized that the implication of calcium-dependent phosphorylation in generating the learning-induced reduction of K^+ currents does not rule out involvement of other biochemical reactions in this process. This is one type of locus, within the cell's biochemical pathways, that a number of experiments indicate is important for biophysical changes lasting at least for a few days. There may be other loci that are affected as a consequence. These may be altered for the same time intervals as calcium-dependent phosphorylation or they may ultimately extend into new temporal domains. It is not inconceivable, for example, that phosphorylation reactions, initially triggered by calcium and lipid mobilization, regulate channel modifications for hours to days and that other reactions (such as those affecting the actual structure of the neuron) regulate modifications for weeks and longer (Figure 65). And whereas the molecular regulation of ionic channels during learning demonstrated thus far involves the cytosolic and membrane compartments of the neuron, participation by the nuclear compartment (containing DNA) in such regulation should not be excluded, particularly for the more or less permanent information storage of long-term memory. Recent experiments, for example, show that activation of Ca^{2+}-dependent phosphorylating pathways in *Hermissenda* modify the rates of synthesis of specific proteins. Modifying the rate of protein synthesis has also been found to produce persistent changes of I_A and $I_{Ca^{2+}-K^+}$ in the *Hermissenda* type B cell.

At the level of the molecular regulation of ionic channels, mechanisms of associative learning, as understood to date for *Hermissenda*, appear to be quite distinct from mechanisms proposed to underly examples of non-associative learning, habituation and sensitization, as studied, for example, in *Aplysia*. Sufficient stimulation of a single sensory pathway is thought to cause increased release of the neurotransmitter serotonin onto sensory cells. During sensitization, serotonin released (from as yet un-

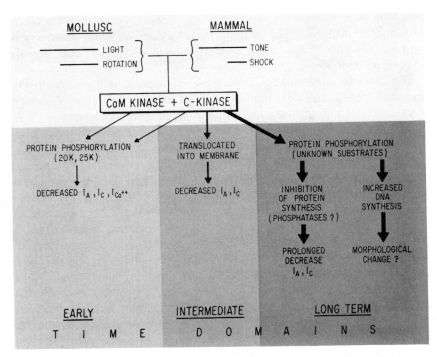

Figure 65. Distinct temporal domains of memory may have corresponding molecular mechanisms and substrates. The immediate effects of associated stimuli involve short-term activation of protein kinases that reduce ionic conductances during acquisition. Intermediate retention may involve a stable translocation of protein kinase from cytosol to membrane. Such translocation greatly prolongs the initial modification of ionic conductances. Even longer-term changes could involve inhibition and/or activation of protein synthesis by protein kinase C (or other kinases such as the calcium–calmodulin-dependent kinase), or modified DNA synthesis. These latter changes, via modification of structure and/or enzymatic proteins, might be responsible for permanent transformations of neuronal excitability.

identified cells), according to this model, activates adenylate cyclase to elevate cyclic-AMP, which in turn stimulates phosphorylation to bring about reduction of a K^+ current called the "S" (or serotonin sensitive) current. Within this model it is not specified whether the S current remains reduced on days during which sensitization persists, whether cyclic-AMP remains elevated, or whether serotonin released from pre-synaptic cells remains increased. The available evidence does indicate, however, that the *Aplysia* S channels, unlike the *Hermissenda* K^+ channels modified by conditioning, are neither voltage-dependent nor sensitive to levels of intracellular calcium. Finally, cumulative depolarization, crucial for encoding stimulus pairings during *Hermissenda* conditioning (see also Chapter 15), is not involved in either habituation or sensitization of *Aplysia*.

Our discussion of molecular regulation of membrane channels during learning would not be complete without some mention of how the association of stimuli used for training is achieved biochemically. It is clear that the distinct visual and vestibular stimuli meet at neuronal convergence points within the neural systems. But, how does this meeting occur within the cell at biochemical convergence points? And how is the requirement for critical timing of the associated stimuli represented in molecular terms? An example of how this occurs has already been suggested. Associated activation of the Ca^{2+}–CaM-kinase pathway and the C-kinase (Ca^{2+}–lipid) pathways produces sustained reduction of I_A and $I_{Ca^{2+}-K^+}$ as well as sustained endocrine responses. Certain protein substrates are phosphorylated by both of these kinases. These shared substrates could be the biochemical convergence points responsible for the prolonged K^+ current reduction of learning (as well as endocrine effects). Unless the initiating elevation of Ca^{2+} within the cell cytoplasm occurs with the proper timing (i.e., after) in relation to lipid availability and unless subsequent sustained flow of Ca^{2+} ions occurs across the cell membrane, maximum and prolonged phosphorylation (and thus K^+ current reduction) may not result (Fig. 66).

For the *Hermissenda* type B cell, the initial Ca^{2+} elevation is produced by the light-elicited depolarization activating the voltage-dependent Ca^{2+} current. Subsequent sustained Ca^{2+} flux, via this current, depends on depolarization due to light *and* to the synaptic excitation triggered by rotation paired with light. Stimulation (usually by a molecule called diacylglycerol), necessary for C-kinase activation, may arise in two ways. First, light itself may be responsible for some lipid release, and second, the synaptic feedback may have such an effect. A number of observations suggest that synaptic excitation of the *Hermissenda* B cell due to light–rotation pairing is substantially effected by synaptic transmission mediated by a neurochemical(s) similar to, or identical with, agents termed as "adrenergic." There is ample precedent in other physiologic contexts for adrenergic agents such as norepinephrine, via its effect on specific membrane receptors, to cause C-kinase activation. Other observations indicate that synaptic inhibition of the *Hermissenda* B cell is effected by the neurochemical acetylcholine. Adrenergic agents decrease type B K^+ currents (I_A and I_C) while acetylcholine increases

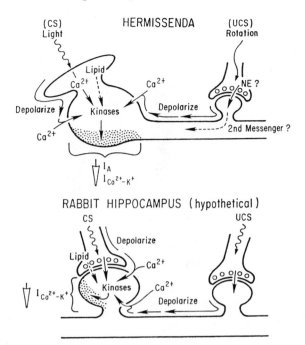

Figure 66. Schematic summary for cellular interaction of conditioned and unconditioned stimuli in *Hermissenda* and hippocampus. (From Alkon, 1986)

these currents. Such neurochemical contributions to the phosphorylation control of ionic currents may have a role, therefore, during *Hermissenda* conditioning.

Although synergistic interaction of Ca^{2+}- and lipid-dependent phosphorylation of cellular proteins may be important for initiating long-lasting reduction of I_A and $I_{Ca^{2+}-K^+}$, we are still confronted with the question of what molecular transformations persist long enough to maintain this reduction for days or longer. If retention of a learned association at a behavior level is translated into K^+ current reduction at the membrane level, how is retention translated at the molecular level? And how does this molecular translation survive cell turnover? A dramatic clue for the solution of this mystery has recently been provided by experiments exploiting the brain slice technique. At least 24 h or longer after classically conditioning (or control procedures), regions of the rabbit brain are sliced and specific portions, even specific neurons, are isolated for biochemical analysis. One such analysis mea-

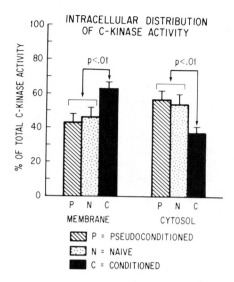

Figure 67. The effect of rabbit eyelid conditioning on the intracellular distribution of protein kinase C in the CA1 region of the hippocampus. The percentage of protein kinase C activity is significantly higher in the membrane fraction of conditioned animals as compared to controls, and significantly lower in the cytosol fraction ($p < 0.01$ ANOVA). Total kinase activity is the same for all groups (conditioned, pseudoconditioned, and naive), suggesting that the C-kinase has been translocated from the cytosol to the membrane 24 h after conditioning. (After Bank et al., 1987)

sures how much of the C-kinase is located in the neurons' cytoplasm and in its outer membrane. This distribution is particularly important because membrane C-kinase is far more active than that in the cytoplasm and is stimulated by very low levels of intracellular Ca^{2+}. Classical conditioning of the rabbit nictitating membrane was found to cause a marked shift of C-kinase distribution (Figure 67). In the hippocampal CA1 neurons (the same cells for which $I_{Ca^{2+}-K^+}$ was reduced; see Chapter 16), *membrane* C-kinase remains increased for more than 24 h after conditioning (but not control procedures). Redistribution of C-kinase, therefore, may provide a molecular storage mechanism that ultimately is responsible for K^+ current reduction for several days.

Other experiments suggest that a similar translocation of C-kinase from cytosolic to membrane compartments could underly biophysical changes for at least hours after *Hermissenda* classical conditioning.

It is possible that conditioning-specific C-kinase translocation in both *Hermissenda* and the rabbit (lasting many hours and even days) ultimately results in modification of DNA- (and/or m-RNA) directed synthesis of critical cell proteins. Activation of C-kinase (via translocation into the membrane) was found to significantly modify protein synthesis as well as m-RNA turnover in *Hermissenda*. Most importantly, recently Dr. Thomas Nelson in our laboratory was able to unequivocally demonstrate conditioning-specific increases of m-RNA turnover in the *Hermissenda* eyes. These increases reach a maximum approximately 1 day after conditioning, but persist at least 3 days longer, and are remarkably closely correlated with the extent of classical conditioning of the intact animal. Recombinant techniques should help reveal which proteins undergo modified synthesis due to these m-RNA changes. Thus, the same medial Type B cell which undergoes prolonged depolarization in the earliest temporal domain would, in progressively longer-lasting temporal domains, show conditioning-specific modification of membrane channels accompanied by C-kinase translocation, changes of RNA metabolism, and ultimately a new balance of protein synthesis and degradation. Such a sequence of cellular transformations would offer a means for transition from the briefest temporal domains (milliseconds-seconds for an initial presentation of stimulus relationships) to those of relative permanence (weeks-years) necessary to account for long-term memory (Figure 65).

Bibliography

Acosta-Urquidi, J., Alkon, D. L., and Neary, J. T. (1984). Ca^{2+}-dependent protein kinase injection in a photoreceptor mimics biophysical effects of associative learning. *Science* 224:1254–7.

Akers, R. F., Lovinger, D. M., Colley, P. A., Linden, D. J., and Routtenberg, A. (1986). Translocation of protein kinase C activity may mediate hippocampal long-term potentiation. *Science* 231:587–9.

Alkon, D. L. (1983). Learning in a marine snail. *Sci. Am.* 249:70–84.

(1984). Calcium-mediated reduction of ionic currents: a biophysical memory trace. *Science* 266:1037–45.

(1986). Conditioning-specific modification of post-synaptic membrane currents in mollusc and mammal. In *The Neural and Molecular Bases of Learning*, ed. by J.-P. Changeux and M. Konishi. Springer-Verlag, New York. In press.

Alkon, D. L., Acosta-Urquidi, J., Olds, J., Kuzma, G., and Neary, J. T. (1983). Protein kinase injection reduces voltage-dependent potassium currents. *Science* 219:303–6.

Alkon, D. L., Kubota, M., Neary, J. T., Naito, S., Coulter, D., and Rasmussen, H. (1986). C-kinase activation prolongs Ca^{2+}-dependent inactivation of K^+ currents. *Biochem. Biophys. Res. Commun.* 134:1245–53.

Bank, B., Coulter, D., Rasmussen, H., Chute, D. L., and Alkon, D. L. (1986). Effects of NMR conditioning on intracellular distribution of protein kinase C. *Soc. Neurosci. Abstr.* 12:182.

Camardo, J. S., Siegelbaum, A. S., and Kandel, E. R. (1984). Cellular and molecular correlates of sensitization in *Aplysia* and their implications for associative learning. In *Primary Neural Substrates of Learning and Behavioral Change*, ed. by D. L. Alkon and J. Farley, pp. 185–204. Cambridge University Press.

Crick, F. (1984). Memory and molecular turnover. *Nature* 312:101.

Flexner, J. B., and Flexner, L. B. (1969). Studies on memory: evidence for a widespread memory trace in the neocortex after the suppression of recent memory by puromycin. *Proc. Natl. Acad. Sci. U.S.A.* 62:729–32.

Hyden, H., and Egyhazi, E. (1962). Nuclear RNA changes of nerve cells during a learning experiment in rats. *Proc. Natl. Acad. Sci. U.S.A.* 48:1366–73.

   (1964). Changes in RNA content and base composition in cortical neurons of rats in a learning experiment involving transfer of handedness. *Proc. Natl. Acad. Sci. U.S.A.* 52:1030–5.

Kaibuchi, K., Takai, Y., Sawamura, M., Hoshijima, M., Fujikura, T., and Nishizuka, Y. (1983). Synergistic functions of protein phosphorylation and calcium mobilization in platelet activation. *J. Biol. Chem.* 258:6701–4.

Kandel, E. R., and Schwartz, J. H. (1982). Molecular biology of learning: modulation of transmitter release. *Science* 218:433–43.

Lisman, J. E. (1985). A mechanism for memory storage insensitive to molecular turnover: a bistable autophosphorylating kinase. *Proc. Natl. Acad. Sci. U.S.A.* 82:3055–7.

Neary, J. T., Crow, T. J., and Alkon, D. L. (1981). Change in a specific phosphoprotein band following associative learning in *Hermissenda*. *Nature* 293:658–60.

Rasmussen, H. (1981). *Calcium and cAMP as Synarchic Messengers*. Wiley, New York.

   (1986a). The calcium messenger system (part I). *N. Engl. J. Med.* 314:1094–1101.

   (1986b). The calcium messenger system (part II). *N. Engl. J. Med.* 314:1164–70.

Rasmussen, H., and Barrett, P. Q. (1984). Calcium messenger system: an integrated view. *Physiol. Rev.* 64:938–84.

Sakakibara, M., Alkon, D. L., DeLorenzo, R., Goldenring, J. R., Neary, J. T., and Heldman, E. (1986a). Modulation of calcium-mediated inactivation of ionic currents by Ca^{2+}/calmodulin-dependent protein kinase II. *Biophys. J.* 50:319–27.

Sakakibara, M., Alkon, D. L., Neary, J. T., Heldman, E., and Gould, R. (1986b). Inositol trisphosphate regulation of photoreceptor membrane currents. *Biophys. J.* 50:797–803.

Sakakibara, M., Collin, C., Kuzirian, A., Alkon, D. L., Heldman, E., Naito, S., and Lederhendler, I. (1987). Effects of α_2-Adrenergic Agonists and Antagonists on Photoreceptor Membrane Currents. *J. Neurochem.* 48:405–416.

Shashoua, V. F. and Holmquist, B. (1986). Extracellular fluid proteins of goldfish brain: evidence for the presence of proteases and esterases. *J. Neurochem.* 47:738–43.

Wood, S. F., Reid, M. S., Szuts, E. Z., and Fein, A. (1986). Rapid formation of inositol trisphosphate ($InsP_3$) in squid photoreceptors. *Biophys. J. Abstr.* 49:30a.

13
A new network bias

The trace left in the type B cell by repeated stimulus pairings is most likely not the only one within the *Hermissenda* visual system. The consequences of the trace in the type B cell are certainly sufficient to account for much of the observed differences in light-elicited signals within the remainder of the visual system (such as from "output" cells or motorneurons). A contribution to these differences apparently also is made by intrinsic membrane changes in at least one other neuron, the type A photoreceptor. Whereas the type B photoreceptor becomes more excitable with the learning of the light–rotation association, type A photoreceptors become less excitable. Whereas I_A and $I_{Ca^{2+}-K^+}$ of the type B cell remain reduced after the conditioning paradigm, they are apparently increased in the type A cell. Thus, the learning-induced biophysical modifications within the two cell types are complementary. They are also complementary in their effects on information flow, at least in one part of the visual pathway. The medial type A photoreceptor causes synaptic excitation of interneurons, which in turn cause synaptic excitation of motorneurons whose impulse activity can cause turning movements of the animal in relation to visual stimuli. The medial type B photoreceptor, on the other hand, inhibits the medial type A cell and thereby inhibits interneurons, motorneurons, and turning in relation to visual stimuli (Figure 68). Both conditioning-induced enhancement of type B excitability and conditioning-induced reduction of type A excitability have a similar effect: They reduce turning movement in response to a visual stimulus. In essence, the conditioning experience has introduced a bias in the visual network against the A cell and in favor of the B cell, a bias that helps determine a change of responses to subsequent presentations of the conditioned stimulus, light.

Behavioral expression of this new network bias does not only depend on the chain of visual cells just described – a chain involved in controlling the direction of visually guided movement. We know that as a result

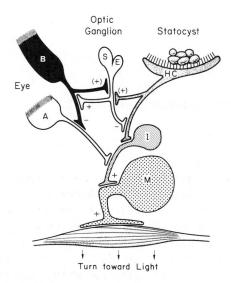

Figure 68. (A) Schematic diagram of a visual pathway and its convergence with the statocyst pathway. The type B photoreceptor (B) causes monosynaptic inhibition of the medial type A photoreceptor (A). The medial type A photoreceptor causes monosynaptic excitation of ipsilateral interneurons (I), which are also excited by ipsilateral hair cells (HC). The ipsilateral interneuron in turn excites the motorneuron (M). Ipsilateral hair cell impulses and type B impulses cause a transient inhibition (not shown here) and are followed by long-lasting effective excitation (+) of the E optic ganglion cell and thereby the type B cell. (From Goh and Alkon, 1984)

of conditioning, light elicits a new behavioral response – foot contraction – that closely resembles the old reflexive response to rotation. We also know that this new conditioned response is closely related to, if not entirely causal of, movement inhibition. The degree of foot contraction predicts the degree to which the velocity of the animal's forward movement is slowed or stopped. Because enhanced type B excitability results in enhanced foot contraction, a visual pathway excited by type B cell synaptic input should be responsible for the movement inhibition during foot contraction. In fact, such a pathway has been found within the *Hermissenda* nervous system. In this pathway, interneurons receive synaptic excitation from the type B cells (in contrast to the first pathway described in which interneurons only receive synaptic excitation from the type A cells).

A conditioning-induced bias in the synaptic network of the *Hermissenda* eye (i.e., in the *relative* excitability of the type A and type B photoreceptors) can create a new balance of behavioral responses to visual stimuli. Again, this new balance can result from complementary shifts of responsiveness in distinct limbs of the visual system. Conditioning causes decreased turning movement toward a light and increased foot contraction (and thus inhibition of movement). Both consequences of conditioning contribute to the decrease of overall movement toward a light source (i.e., a reduction of phototactic responsiveness).

The conditioning-induced bias within the visual network, together with the resulting shift in the balance of behavioral responses to visual stimuli, must be considered for an understanding of another behavioral feature of conditioning, extinction, on a cellular level. During extinction, repeated presentations of the conditioned stimulus (light) alone during the period of memory retention reduces the duration of that retention. Repeated conditioned stimulus presentation extinguishes the learned association of light and rotation. Why? After all, light causes depolarization of the type B cell, increased input resistance, and reduced I_A and $I_{Ca^{2+}-K^+}$. Why should not more light cause more of these type B cell changes and thus further acquisition of the learned stimulus association?

First, there are features of the type B photoreceptor itself that limit the amount of depolarization that light presentations alone can produce. Too much light will bleach the visual pigment of the photoreceptor. This pigment, rhodopsin (also present in our own photoreceptors), is necessary for transducing light energy into electrical signals. Light changes the rhodopsin molecule and thereby initiates a biochemical sequence that results in the opening of type B sodium channels. Sodium flows from the outside to the inside of the cell thereby causing it to depolarize. Once the rhodopsin molecules have been affected by light, time is necessary before they once more become sensitive to light. This time is called the period of dark adaptation. Thus, the first depolarization of a dark-adapted cell may be quite large, but without sufficient time for the type B cell to return to its former level of dark adaptation, depolarization of the cell in response to a second presentation of light may be much smaller than the first. The time necessary to keep the type B cell somewhat dark-adapted (i.e., sensitive to light stimuli) limits the amount of depolarization that can result from frequent presentation of light alone.

Another feature of the type B photoreceptor itself that constrains its sensitivity is the dependence of the light-induced Na^+ current on the level of intracellular Ca^{2+}. Elevated intracellular Ca^{2+} inactivates the light-induced flux of Na^+ across the type B rhabdomeric (light-sensitive) membrane. This inactivation is quite different from Ca^{2+}-mediated inactivation of K^+ currents (during conditioning). It does not last for days, nor does it result from the same biochemical manipulations (see Chapter 12) as does K^+ current inactivation. K^+ current inactivation results from intracellular injection of the enzyme Ca^{2+}–calmodulin-dependent kinase or an intracellular molecular messenger, called inositol-tris-phosphate, whereas the light-induced current of the type B cell is not affected. Similarly, activation of the C-kinase with a phorbol ester followed by a Ca^{2+} load causes K^+ current inactivation but not inactivation of the light-induced Na^+ current. Conversely, we have shown experimentally that injection of Mg^{2+} into the cell inactivates the light-induced Na^+ current without affecting the K^+ currents. Ca^{2+}-mediated inactivation of the Na^+ current, nevertheless, can be quite substantial. Depolarization of the type B cell during its response to light activates the voltage-dependent flux of Ca^{2+} across the type B soma membrane. The resulting elevation of Ca^{2+} temporarily inactivates the light-induced Na^+ current as well as K^+ currents. Inactivation of Na^+ current reduces depolarization of the type B cell in response to subsequent light stimuli. Because, over long time periods such as those of acquisition or retention of the learned association, inactivation of Na^+ current does not last as long as inactivation of K^+ current, a sufficient interval of time between light–rotation presentations will allow more recovery from Na^+ current inactivation than K^+ current inactivation. Again, the point is that, for a maximal increase of type B excitability too frequent or too bright, light stimuli will be ineffective.

But too much light is ineffective in producing a long-term change of type B excitability for another reason – one involving the shift of the relative excitability of the type A cell and type B cell excitability during conditioning. Light alone stimulates both type A and type B photoreceptors, whereas light paired with rotation – even one such pairing – causes more depolarization of the type B cell and less depolarization of the type A cell than occurs with light alone. Each pairing augments the *de*polarization of the type B cell and the *hyper*polarization of the type A cell a bit more (Figure 69). Prior to training, the type A cell hyperpolarizes (i.e., its membrane potential becomes more negative than the resting level) after a light stimulus, whereas the type B cell depolarizes. After light paired with rotation, the type B cell depolarization increases, whereas the type A cell hyperpolarization increases. Every presentation of light alone acts to restore the original balance of type A cell and type B cell excitability. The more that depolarization of the type B cell results from light alone (i.e., due to its brightness or frequency), the more closely does type A depolarization approximate that of the type B. Light alone

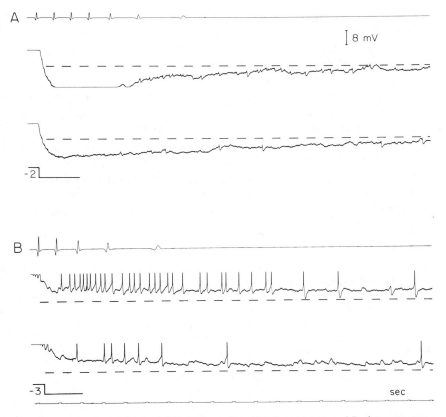

Figure 69. The effect of paired light and rotation stimuli on type A and B photoreceptor responses. (A): Responses of intact lateral type A photoreceptor to light alone (lower record) and light paired with rotation (upper record). The end of a 10-s light step (intensity in −log units) is indicated by the bottom trace. The end of a rotation stimulus (onset 5 s before light), with a maximum of 1.3 g, is indicated by the top trace. The number of IPSPs and hyperpolarization of the type A cell *after* the light step are shown. Note that the recording is saturated immediately after light when paired with rotation. (B): Responses of intact type B photoreceptor to light alone (lower record) and light paired with rotation. The end of a 20-s light step (intensity in −log units) is indicated by the bottom trace. Rotation, indicated by the top trace, is as above. The number of impulses during the LLD is greatly increased by paired stimulation. (From Alkon and Grossman, 1978)

reduces the difference between type A cell and B cell depolarization and excitability, whereas light paired with rotation enhances that difference.

This, then, may be an important translation of the behavioral phenomenon of extinction to the cellular level. The interaction of individual features of the type A cells and the type B cells with features of the visual–statocyst network's synaptic responses to light and rotation stimuli dictate that unpaired presentations of the conditioned stimulus (light) alone will degrade the network bias necessary for encoding and recalling the learned association of light and rotation. Just as unpaired presentations of the light alone will extinguish the learned behavior (i.e., when light elicits a new, conditioned response), such unpaired presentations will also extinguish the learning-induced increase of type B excitability and the complementary learning-induced decrease of type A excitability.

Bibliography

Alkon, D. L., and Grossman, Y. (1978). Long-lasting depolarization and hyperpolarization in eye of *Hermissenda*. *J. Neurophysiol.* 41:1328–42.

Alkon, D. L., Shoukimas, J., and Heldman, E. (1982). Calcium-mediated decrease of a voltage-dependent potassium current. *Biophys. J.* 40:245–50.

Alkon, D. L., and Sakakibara, M. (1985). Calcium activates and inactivates a photoreceptor soma potassium current. *Biophys. J.* 48:983–95.

Farley, J. (1986). Contingency learning and causal detection in *Hermissenda*: behavioral and cellular mechanisms. *Behav. Neurosci.* In press.

Farley, J., Richards, W., and Alkon, D. L. (1982). Extinction of associative conditioning in *Hermissenda*: behavior and neural correlates. *Bull. Psychonom. Soc.* 20:144.

Goh, Y., and Alkon, D. L. (1984). Sensory, interneuronal and motor interactions within the *Hermissenda* visual pathway. *J. Neurophysiol.* 52:156–69.

Richards, W. G., Farley, J., and Alkon, D. L. (1984). Extinction of associative learning in *Hermissenda*: behavior and neural correlates. *Behav. Brain Res.* 14:161–70.

Sakakibara, M., Alkon, D. L., DeLorenzo, R., Goldenring, J. R., Neary, J. T., and Heldman, E. (1986a). Modulation of calcium-mediated inactivation of ionic currents by Ca^{2+}/calmodulin-dependent protein kinase II. *Biophys. J.* 50:319–27.

Sakakibara, M., Alkon, D. L., Neary, J. T., Heldman, E., and Gould, R. (1986b). Inositol trisphosphate regulation of photoreceptor membrane currents. *Biophys. J.* 50:797–803.

14

Molluscan versus mammalian brain

All of what we have learned from cellular mechanisms of learning in *Hermissenda* indicates that in order for two stimuli to be associated, they must independently initiate a chain of electrophysiologic signals that meet somewhere in the nervous system. In order for the animal to learn that one stimulus is related to another stimulus – that one stimulus predicts the occurrence of the other stimulus – the two stimuli must first affect the same neuronal structure, that is, the two stimuli must affect neural pathways that converge. Convergence is determined by the genetically programmed neuron networks that are already formed – that exist prior to the occurrence of the stimulus relationship to be learned. Light stimulates the type B photoreceptor directly and rotation stimulates the same cell indirectly (i.e., via synaptic input). The type B photoreceptor is, therefore, a site for convergence, right at the input stage of the visual pathway. Interneurons in the visual pathway receive synaptic excitation from both the medial type A photoreceptor and hair cells. Such interneurons are also convergence sites but at the first synaptic junctions within the *Hermissenda* visual and statocyst pathways.

Convergence in the mammalian brain rarely occurs at the input stage of sensory pathways. Auditory sensory cells, for example, are stimulated directly by sound waves and indirectly by synaptic signals received from neurons in the central nervous system. Vertebrate photoreceptors, however, do not receive input from other sensory modalities. In fact, visual signals usually do not converge with other sensory signals until they reach groups of neurons located separately from the retina, that is, after the signals have serially activated a sequence of synaptic junctions. The vast number of convergence sites in the mammalian brain are in the central nervous system, in marked contrast to the peripheral location of the visual–statocyst convergence in *Hermissenda* (as well as other molluscs).

Yet the way converging signals evoke progressive transformation of neuronal excitability might be strikingly similar throughout a wide range of animal species. This possibility was suggested some years ago by some of the then known properties common to both the molluscan and mammalian nervous systems.

The hippocampal pyramidal cell, for example, has ionic channels within the membrane of its cell body, or soma, quite similar to those of the *Hermissenda* type B photoreceptor. In fact, in many respects, the pyramidal cell could be regarded as a type B cell transplanted into the mammalian hippocampus. Both cell types have a rapidly activating and inactivating inward sodium current. They both have a sustained inward Ca^{2+} current that is activated by positive shifts of membrane potential. There are, in addition, three outward K^+ currents common to both cell types: I_A, $I_{Ca^{2+}-K^+}$, and a delayed current called I_{K^+}. Furthermore, it has been possible to show that stimulation of synaptic inputs to the pyramidal cell causes marked depolarization and prolonged elevation of intracellular Ca^{2+}. Some, but not necessarily all, of these features may also occur at sites that receive the majority of synaptic input – the dendrites. Vertebrate neurons, unlike those of molluscs, have very elaborate branches that extend from the cell body. Still smaller branches, called dendrites, extend from the larger branches. Because of their size our knowledge of electrophysiologic dendrite properties is still quite limited, but what we do know suggests that Ca^{2+} elevation and some of the ionic channels observed in cell somata can also occur on the dendrites. By far the most common site for convergence of inputs is the dendrite. This conclusion is based on the microscopic measurements of the distribution of anatomically identifiable presynaptic endings along the structures of mammalian central neurons. So, we might expect to find conditioning-induced changes of membrane channels on or near dendrites more frequently than on the cell bodies and much more frequently than on axons or axonal endings (presynaptic terminals).

The properties common to molluscan and mammalian neurons suggested the potential for common learning mechanisms. But how could such mechanisms be established? As discussed earlier, clinical experience, lesion studies, and our intuition suggest that learned stimulus associations are stored in mammals in a distributed fashion. No one cell, or restricted area of cells, in a vertebrate brain would be expected to have exclusivity over the representation of a particular memory. Certainly some cells and/or cell areas can be expected to be more crucial for specific aspects of the memory – for instance, involvement of discrete movements, emotional context, spatial orientation, etc. But, given the vastly increased resolution of sensory discrimination and motor control of vertebrate species, we would not expect that changes within only a few neurons can (as they do in *Hermissenda*) bear a causal relationship to the acquisition of a learned association. What we would expect are sets of conditioning-specific changes within several brain areas, which together have causal impact. We would also expect that such changes, if they could be found, should be intrinsic to the neurons themselves. If these neurons could be removed from a mammal on days after it had been conditioned, conditioning-specific modifications would still be present, even in the absence of synaptic contact with other brain areas and without neurohumoral influences borne by the circulating fluids in the living animal. An approximation of the necessary experimental conditions was provided by the brain slice technique. Thin slices (e.g., 200–300 μm) were made from a slab removed from a restricted brain region, and the biophysical properties of neurons within the slice are then assessed under artificial conditions that are as physiologic as possible (Figure 70). But where were these slices to be made? What previous studies might guide the investigation and suggest where associatively learned information might be stored?

Previously, an increase of impulse activity elicited by a conditioned stimulus had been correlated with the acquisition and retention by intact animals of a learned association. One such association involved classical conditioning of the rabbit nictitating membrane. The nictitating membrane reflexly contracts in response to a puff of air or a mild discrete electric shock to the surface (or cornea) of the eye. When a tone (or light stimulus) repeatedly precedes the air puff, the animal learns to respond to the tone as if it were the air puff. Namely, during classical conditioning the tone (conditioned stimulus) comes to predict the subsequent occurrence of the puff (unconditioned stimulus).

Electrodes places outside of pyramidal cells (see Figures 71 and 72) (i.e., extracellular electrodes) within the rabbit hippocampus record the number of impulses elicited by the tone within a chosen time interval. These cells, called the CA1 pyramidal cells, respond with more impulses (recorded extracellularly), with less delay, to tone presentation as the animal learns the tone–puff association (Figure 73). Did this increase of tone-elicited impulse activity represent a learning-induced change intrinsic to the hippocampus or perhaps even intrinsic to the CA1 pyramidal cells, or was it simply a consequence of learning-specific differences of synaptic input due to changes of neurons in other unknown brain

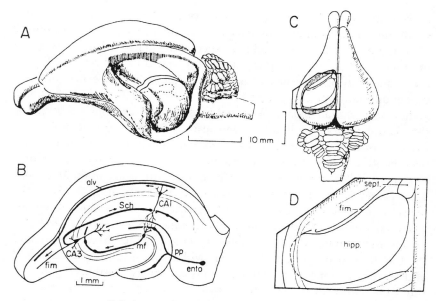

P. Andersen, T.V.P. Bliss and K.K. Skrede, 1971

Figure 70. Anatomy of the hippocampal formation. (A) Lateral view of the rabbit brain with the parietal and temporal neocortex removed to expose the hippocampal formation. The indicated section of the hippocampal formation is enlarged (B) to show the main neuronal elements of the structure. (C) Diagram of the rabbit's brain seen from above with the neocortex removed to expose the left hippocampal formation. The indicated window is enlarged (D) to show the details of the hippocampal formation. Similar diagrams are used in other figures to indicate the orientation of the fiber system under study. alv, alveus; ang. bundle, angular bundle; ento, entorhinal area; f. hipp., hippocampal fissure; fin., finbria; gran., granular layer; hipp., hippocampus; ms, mossy fibers; p.p., perforant path; pyr., pyramidal layer; Rec., recording electrode; Sch, Schaffer collaterals; sept., septum; Stim., stimulating electrode. (From Andersen et al., 1971)

regions? Was learned information actually stored in the CA1 neurons or were these neurons' activity only reflecting information stored elsewhere?

To address this question, slices of the hippocampus were made from conditioned, as well as control, animals at least 24 h after the completion of the training experience. Intracellular recordings were made from CA1 neurons within these slices over the next 10 h to assess intrinsic membrane properties. Many properties (such as resting membrane potential, input resistance, and impulse amplitude) showed no differences between conditioned and control animals. Measurements of one particular feature, however, were markedly different for conditioned, as compared to naive or to "pseudoconditioned" (i.e., receiving unpaired tone and shock

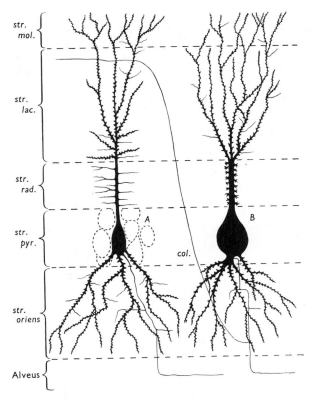

Figure 71. Pyramidal cells of Ammon's Horn, redrawn from Cajal (1911). The regions of the neuron are indicated at the left of the figure. Alveus, white matter containing the efferent axons; str. oriens, region of basal dendrites; str. pyr., region of neuronal perikarya; str. rad., first part of apical dendrite; str. lac., region of main branches of apical dendrite; str. mol., region of final branches of apical dendrites. "A" is a small pyramidal neuron from area CA1. Note the thin lateral branches in the stratum radiatum and the absence of spines. "B" is a large pyramid from area CA3. There are few lateral branches in the stratum radiatum but the main dendritic trunk is characterized by the presence of large spines, which are often branched. Note the long axonal collateral (col) of Schaffer, which ascends to the upper part of the stratum lacunosum of the small pyramids in CA1. (From Hamlyn, 1963)

stimuli), animals. Following a positive current pulse that triggers one or more impulses, the membrane potential of the CA1 pyramidal cell becomes more negative than that of the previous resting level. This "afterhyperpolarization," when measured at least 100 msec after the impulses, has been shown to be due to activation of an outward Ca^{2+}-dependent K^+ current, $I_{Ca^{2+}-K^+}$, the same current that was reduced in the type B cells of conditioned *Hermissenda*. Remarkably, the

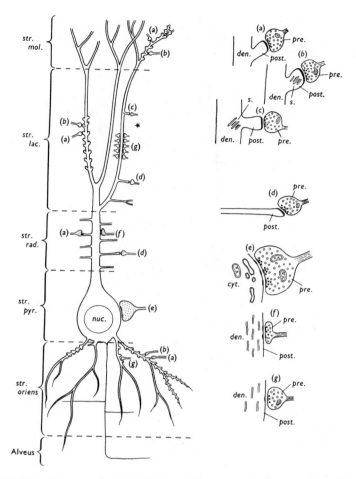

Figure 72. A diagram, not to scale, of a pyramidal neuron from region CA1 of Ammon's Horn, showing the levels at which different varieties of synaptic contact, (a)—(g), have been observed. Details of these types are shown at the right of the figure. For the sake of clarity the different synaptic arrangements in the stratum lacunosum have been shown on three branches. Any of these varities may, of course, be found on any dendritic branch in the region. (From Hamlyn, 1963)

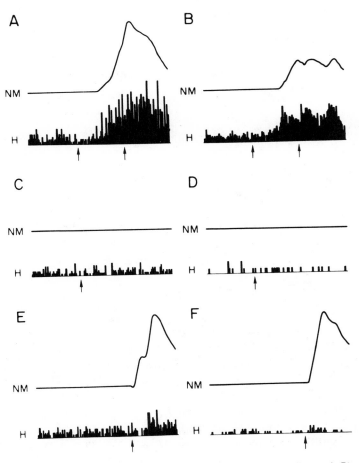

Figure 73. Responses of identified pyramidal neurons during paired, (A) and (B), and unpaired, (C)—(F), presentations of tone and corneal air puff. In this figure, the upper traces show the averaged NM response for all trials during which a given cell was recorded. The bottom traces show the peristimulus time histogram produced by the response of the recorded neuron during the pre-CS, CS, and UCS periods, respectively. The total trial length of both NM responses and histograms is 750 ms. Arrows occurring early in the trial period indicate tone onset; arrows occurring late in the trial period indicate air puff onset. NM, nictitating membrane; H, hippocampus. For this figure, (A) and (B) show examples of responses of two pyramidal neurons recorded from two different animals during paired conditioning. Note that the conditioned NM response of animal in (A) is monophasic, and the histogram of pyramidal cell firing parallels the NM amplitude–time course with a unimodal within trial distribution of action-potential discharges. The conditioned NM response of animal in (B) is triphasic, and again the histogram of pyramidal cell activity parallels the NM response, but with a trimodal distribution. Results in (C) and (E) show response of a pyramidial neuron recorded from an animal given unpaired tone-alone (C) and air puff-alone (E) presentations. (D) and (F) are the same for a different pyramidal cell recorded from a different control animal. (From Berger et al., 1983)

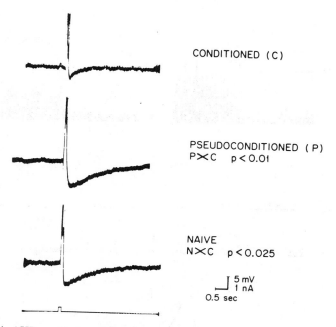

Figure 74. AHP amplitude and duration is reduced by conditioning. Typical AHP traces for conditioned, pseudoconditioned, and naive cells following a 100-ms depolarizing current pulse eliciting four spikes. Calibration: 5 mV, 0.5 s. (From Coulter et al., 1987)

afterhyperpolarization, and thus the $I_{Ca^{2+}-K^+}$, of conditioned CA1 neurons was reduced in amplitude and duration as compared to naive and pseudoconditioned animals (Figures 74 and 75). This $I_{Ca^{2+}-K^+}$ reduction, intrinsic to the CA1 neurons, could explain the conditioning-specific increase of CA1 excitability (manifested by the increased impulse frequency elicited with less delay by the conditioned stimulus). Signals will depolarize the CA1 cell more readily because less $I_{Ca^{2+}-K^+}$ (which opposes depolarization) will be activated in conditioned cells. Does this mean *all* synaptic inputs to the CA1 cell will more readily trigger impulses in conditioned animals? Not at all. Why, requires mention of a few more observations, and consideration of the possible locus on the pyramidal cell structure of the $I_{Ca^{2+}-K^+}$ reduction.

It is important to remember that our intracellular microelectrode is recording the afterhyperpolarization in the soma of the CA1 cell. The reduction of afterhyperpolarization, however, is apparent only following one or more impulses – impulses that travel very quickly to other parts of the CA1 geometry – particularly the dendrites. Depolarization of the CA1 neuron to potential levels below the threshold for impulse activity

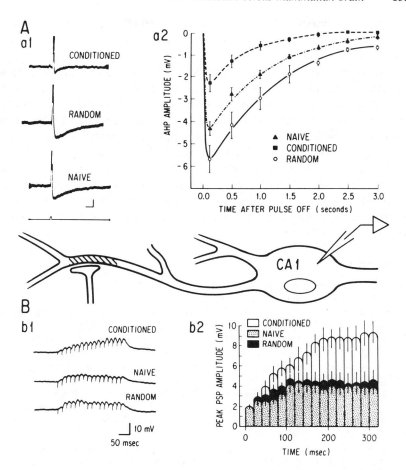

Figure 75. Conditioning reduces the AHP and enhances the summation of synaptic potentials. A) Representative records (a1) and plot of means ± standard errors (a2) of AHPs for 19 cells from conditional animals (■), 18 cells from pseudoconditional animals (○), and 19 cells from naive animals (▲). B) Representative records (b1) and plot of means ± standard errors (b2) of synaptic potentials for 15 cells from conditional animals (□), 15 cells from pseudoconditional (random) animals (■), and 15 cells from naive animals (□).

causes small afterhyperpolarizations that are not different for conditioned, as compared to control, animals. Furthermore, the difference of afterhyperpolarization between groups increases with increasing number of preceding impulses. So what could be happening is that positive current injected into the cell soma causes impulses that travel (by regenerative propagation) to dendrites where a significant afterhyperpolarization follows the impulses. This afterhyperpolarization then travels (by nonregenerative or passive spread) back to the soma to be recorded by the microelectrode.

Consistent with this possibility was the observation that the input resistance, namely, the voltage change resulting from a constant current injection, was not different for conditioned as compared to control groups. This measure of CA1 excitability largely reflects the soma membrane properties. Thus, conditioning may result in reduction of $I_{Ca^{2+}-K^+}$ across the membrane of particular dendrites or groups of dendrites rather than the membrane of the soma. Consistent with such a dendritic locus, a conditioning-specific change of synaptic potential summation (which occurs on the dendrites) was recently found to be correlated both with the conditioned behavioral response and the change of afterhyperpolarization (Fig. 75).

Bibliography

Alkon, D. L. (1984). Calcium-mediated reduction of ionic currents: biophysical memory trace. *Science* 226:1037–45.

Andersen, P., Bliss, T. V. P., and Skrede, K. K. (1971). Lamellar organization at hippocompal excitatory pathways. *Exp. Brain Res.* 13:222–38.

Berger, T. W., Rinaldi, P. C., Weisz, D. J., and Thompson, R. F. (1983). Single-unit analysis of different hippocampal cell types during classical conditioning of rabbit nictitating membrane response. *J. Neurophysiol.* 50:1197–1219.

Coulter, D. A., Kubota, M., Disterhoft, J. F., Moore, J. W., and Alkon, D. L. (1987). Classical conditioning alters the amplitude and time course of the calcium-dependent afterhyperpolarization in rabbit hippocampal pyramidal cells. *J. Neurophysiol.* In press.

Disterhoft, J. F., Coulter, D. A., and Alkon, D. L. (1986). Conditioning-specific membrane changes of rabbit hippocampal neurons measured in vitro. *Proc. Natl. Acad. Sci. U.S.A.* 83:2733–7.

Gobel. S., Falls, W. M., Bennett, G. J., Abdelmoumene, M., Hyashi, H., and Humphrey, E. (1980). An EM analysis of the synaptic connections of horseradish peroxidase-filled stalked cells and islet cells in the substantia gelatinosa of adult cat spinal cord. *J. Comp. Neurol.* 194:781–807.

Hamlyn, L. H. (1963). An electron microscope study of pyramidal neurons in the Ammon's Horn of the rabbit. *J. Anat. (Lond.)* 97:189–201.

Lo Turco, J. J., Coulter, D. A., and Alkon, D. L. (1987). Enhanced summation of synaptic potentials: a correlate of associative memory in rabbit CA1 pyramidal neurons. *Science*, in press.

Mishkin, M., Ungerleider, L. G., and Macko, K. A. (1983). Object vision and spatial vision: two cortical pathways. Trends in Neurosciences, 10, 414–17.

Pellionisz, A., and Llinas, R. (1979). Brain modeling by tensor network theory and computer simulation. The cerebellum: distributed processor for predictive coordination. *Neuroscience* 4:323–48.

15
Other models and observations

Among a number of preparations (Tables 1 and 2) currently being used to study learning physiology, six, which have been analyzed in sufficient detail to propose or explore cellular mechanisms, are:

1. classical conditioning of the mollusc *Hermissenda*;
2. eyeblink and nictitating membrane conditioning in cats and rabbits;
3. sensitization of the mollusc *Aplysia*;
4. long-term potentiation in the hippocampus;
5. long-term depression in the cerebellum; and
6. classical conditioning of cat paw withdrawal.

Hermissenda, rabbit hippocampus, and Aplysia

As developed in the previous chapters, classical conditioning of *Hermissenda* was shown to result, at least in part, from reduction of two specific K^+ currents, I_A and $I_{Ca^{2+}-K^+}$. I_A and $I_{Ca^{2+}-K^+}$ reduction was also shows to be *intrinsic* to postsynaptic neuronal membranes (type B somata) and not explainable as a passive reflection of neurohumoral or synaptic input from other neurons. Another instance where the intrinsic nature of a learning-induced change has been examined is the rabbit hippocampus. Here it was possible to find reduction of CA1 afterhyperpolarization (a measure of $I_{Ca^{2+}-K^+}$) intrinsic to slices of classically conditioned rabbits. The means of eliciting the afterhyperpolarization (i.e., by injection of positive current through an intracellular microelectrode within the CA1 cell, (even after the elimination of synaptic transmission by perfusion with the impulse blocker, TTX) indicates that the conditioning-specific $I_{Ca^{2+}-K^+}$ reduction is intrinsic to the CA1 membrane (as it was to the type B soma membrane). Reduction of K^+ current(s) seems to be initially (i.e., for the first few minutes) important for *Aplysia* sensitization. However, it is not yet known whether these K^+ currents are intrinsic membrane changes, whether they are localized in

135

Table 1. *Behavioral features of learning models*

Species	Type of learning	Temporal specificity	CS–UCS transfer	Duration
Aplysia	Sensitization	−	−	Weeks
	conditioning	+	−	Days
Hermissenda	Pavlovian conditioning	+	+	Weeks
Rabbit	Pavlovian conditioning	+	+	Months
Cat	Pavlovian conditioning	+	+	Months
Long-term potentiation (LTP)	−	+	−	Days or weeks
Long-term depression (LTD)	Vestibular— ocular adaptation	±	−	1–2 h

Note: Dash means not conclusively demonstrated.

presynaptic terminals, or whether they persist during the retention period for the sensitization. Although the *in vivo* function of serotonin in *Aplysia* learning is still undefined, it does seem that a substance like serotonin, which is one of a class of neurotransmitters called mono-amines, could amplify the effects of sensitizing stimuli. A similar ampli-fying function for a monoamine (adrenergic substance) has also been suggested for *Hermissenda* conditioning (see Chapter 12).

Table 2. *Behavioral features of learning models*

Species	Stimulus specificity	Extinction	Contingency	Savings
Aplysia	−	−	−	−
	+	−	−	−
Hermissenda	+	+	+	+
Rabbit	+	+	+	+
Cat	+	+	+	+
Long-term potential (LTP)	−	−	−	−
Long-term depression (LTD)	+	−	−	−

Note: Dash means not conclusively demonstrated.

Long-term potentiation and long-term depression

Two other models of learning have generated useful information and an interesting direction for future research. Both of these models use electrical stimulation of nerve tracts rather than natural stimulation of neural pathways. Long-term potentiation (LTP) is by definition a long-lasting (up to weeks) enhancement of synaptic transmission induced by high-frequency, called tetanic, stimulation of presynaptic fibers in the hippocampus. The enhanced synaptic transmission is manifest by an increased EPSP amplitude, a decreased latency for EPSP-triggered post-synaptic impulses, and an increased probability of EPSP-triggered impulses. Both pre- and postsynaptic mechanisms have been implicated in the origin of LTP. Prolonged depolarization induced by the neuro-transmitter glutamate together with elevation of intracellular Ca^{2+} appear to be necessary within the postsynaptic dendrites (Figure 76 and Tables 3 and 4). In this respect the initiation steps in *Hermissenda* classical conditioning and LTP are remarkably similar – both involving prolonged depolarization, summation of EPSPs, elevated Ca^{2+}_i, and neurotransmitter effects (an adrenergic substance for *Hermissenda*,

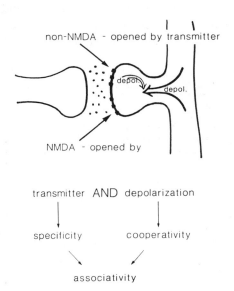

Figure 76. A model for the induction of LTP on a postsynaptic site. [(After Wigstrom and Gustaffson, 1985)]

Table 3. *Cellular features of learning models*

Species	Type of learning	Postsynaptic	Initial depolarization	Initial elevation of Ca^{2+}	Initial K^+ current reduction
Aplysia	Sensitization	−	−	−	+
	conditioning	−	−	−	−
Hermissenda	Pavlovian conditioning	+	+	+	+
Rabbit	Pavlovian conditioning	+	−	−	−
Cat	Pavlovian conditioning	+	−	−	−
Long-term potentiation (LTP)	−	±	±	±	−
Long-term depression (LTD)	Vestibular– ocular adaptation	+	+	+	−

Note: Dash means not conclusively demonstrated.

glutamate for LTP). More unique to LTP are morphologic and biochemical changes that have been reported but have not yet been consistently observed. Whereas swelling of postsynaptic dendritic spines has been reported by some, a lack of swelling, but an increased number of symmetrically arranged presynaptic contacts, have been reported by

Table 4. *Cellular features of learning models*

Species	Persistent K^+ current reduction	Neurotransmitter amplification	C-kinase implicated	Ca^{2+}–CaM dependent kinase implicated	Cyclic-AMP dependent kinase implicated
Aplysia	−	+	−	−	+
	−	−	−	−	−
Hermissenda	+	−	+	+	−
Rabbit	+	−	+	−	−
Cat	+	−	−	+	−
Long-term potentiation (LTP)	−	+	+	−	−
Long-term depression (LTD)	−	+	−	−	−

Note: Dash means not conclusively demonstrated.

others. Whereas one group has reported increased binding of radioactive glutamate on postsynaptic membranes after LTP, other groups have not been able to reproduce this finding. One interesting series of biochemical studies (see Chapter 12) has implicated C-kinase-mediated phosphorylation of neuronal proteins in LTP – again a biochemical step that might be shared with *Hermissenda* conditioning as described earlier. Finally, in addition to postsynaptic mechanisms, presynaptic mechanisms may participate in LTP generation. Evidence for presynaptic involvement includes a demonstration of increased release of glutamate from presynaptic fibers stimulated to produce LTP. Also, statistical analysis of the number of unitary transmitter packets ("quanta") released is consistent with an increased number of packets released at least for the first few hours of LTP (demonstrated, however, only for the crayfish).

Long-term depression (LTD) is, by definition, reduction for approximately 1 h of EPSP amplitude recorded from cerebellar Purkinje cells after concurrent high-frequency stimulation of two distinct types of presynaptic fibers called the parallel fibers and the climbing fibers. Here again, the available evidence suggests that the crucial initial steps are prolonged depolarization, elevation of Ca^{2+}_i, and glutamate activation of postsynaptic dendritic receptors (Tables 3 and 4 and Fig. 77). Although the duration and nature of the resulting neural changes are quite different from LTP and *Hermissenda* classical conditioning, the sharing of initial steps is striking.

Conditioned paw flexion. A classical conditioning paradigm was used to modify cat paw flexion. Thus far, long-term changes (increased amplitude and decreased latency) of EPSPs recorded from neurons, which are a CS–UCS convergence site, have been reported. These neurons, within an extensively analyzed structure called the "red nucleus," were also found to undergo structural changes at points of synaptic contact in response to lesions of critical input tracts to the nucleus. Changes of EPSP size and latency were also correlated with these structural differences and thus the possibility exists that there is some structural basis for the conditioning-induced EPSP alterations.

Conditioned eyeblink of the cat. One of the early electrophysiologic studies of classical conditioning was made with the cat eyeblink preparation (similar to that described for the rabbit). Clear extracellular correlates specific to classically conditioned animals were obtained from motor cortex neurons. Later studies suggested that an increased input

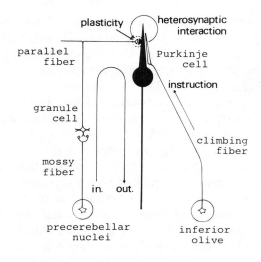

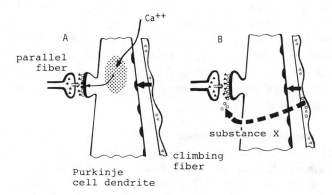

Figure 77. (A) Diagrammatical illustration of the Marr–Albus model. The inhibitory neuron is filled in black. (B) Diagrammatical illustration for the two possible mechanisms of the heterosynaptic interaction between a climbing fiber and a parallel fiber–Purkinje cell synapse. (From Ito, 1982)

resistance might account for a conditioning-specific increase of excitability of these cortical neurons. The cat eyeblink preparation is the only one analyzed thus far *in vivo* with intracellular recording. Increased excitability and increased input resistance were, as previously described, characteristic of type B cells isolated from classically conditioned *Hermissenda*. Finally, also similar to results with *Hermissenda*, injection of Ca^{2+}–CaM-dependent kinase into the cortical cells when coupled with a

Ca^{2+} load produced increased input resistance (albeit a transient one, e.g., 20–25 s, as opposed to hours with *Hermissenda*).

Bibliography

Akers, R., and Routtenberg, A. (1986). Kinase C activity may mediate hippocampal long term potentiation. *Science* 231:587–9.

Alkon, D. L. (1984). Calcium-mediated reduction of ionic currents: biophysical memory trace. *Science* 226:1037–45.

(1985). Changes of membrane currents and calcium-dependent phosphorylation during associative learning. In *Neural Mechanisms of Conditioning*, ed. by D. L. Alkon and C. D. Woody, pp. 3–18. Plenum, New York.

Andersen, P., Sundberg, S. H., Sveen, O., and Wigstrom, H. (1977). Specific long-lasting potentiation of synaptic transmission in hippocampal slices. *Nature* 266:736–7.

Andersen, P., Sundberg, S. H., Sveen, O., Swann, J. N., and Wigstrom, H. (1980). Possible mechanisms for long-lasting potentiation of synaptic transmission in hippocampal slices from guinea pigs. *J. Physiol* 302:463–82.

Bliss, T. V. P., and Lomo, T. (1973). Long-lasting potentiation of synaptic transmission in the dentate area of the anaesthetized rabbit following stimulation of the perforant path. *J. Physiol. (Lond.)* 232:331–56.

Camardo, J. S., Siegelbaum, A. S., and Kandel, E. R. (1984). Cellular and molecular correlates of sensitization in *Aplysia* and their implications for associative learning. In *Primary Neural Substrates of Learning and Behavioral Change*, ed. by D. L. Alkon and J. Farley, pp. 185–204. Cambridge University Press.

Carew, T. J., Abrams, T. W., Hawkins, R. D., and Kandel, E. R. (1984). The use of simple invertebrate systems to explore psychological issues related to associative learning. In *Primary Neural Substrates of Learning and Behavioral Change*, ed. by D. L. Alkon and J. Farley, pp. 169–84. Cambridge University Press.

Cohen, D. H. (1984). Identification of vertebrate neurons modified during learning: analysis of sensory pathways. In *Primary Neural Substrates of Learning and Behavioral Change*, ed. by D. L. Alkon and J. Farley, pp. 129–54. Cambridge University Press.

Disterhoft, J. F., Coulter, D. A., and Alkon, D. L. (1986). Conditioning-specific membrane changes of rabbit hippocampal neurons measured in vitro. *Proc. Natl. Acad. Sci. U.S.A.* 83:2733–7.

Horn, G., Rose, S. P. R., Bateson, P. P. G. (1973). Experience and plasticity in the central nervous system. *Science* 181:506–14.

Ito, M. (1982). Synaptic plasticity underlying the cerebellar motor learning investigated in rabbit's flocculus. In *Conditioning: Representation of Involved Neural Functions*, ed. by C. D. Woody, pp. 213–22. Plenum, New York.

Kandel, E. R. (1976). *Cellular Basis of Behavior: An Introduction to Behavioral Neurobiology*. Freeman, San Francisco.

Kandel, E. R., and Schwartz, J. H. (1982). Molecular biology of learning: modulation of transmitter release. *Science* 218:433–43.

Lomo, T. (1966). Frequency potentiation of excitatory synaptic activity in the dentate area of the hippocampal formation. *Acta Physiol. Scand. (Suppl.)* 68:277, 128.

Lynch, G., Larson, J., Kelso, S., Barrionuevo, G., and Schottler, F. (1983). Intracellular injections of EGTA block induction of hippocampal long-term potentiation. *Nature* 305:719–21.

Thompson, R. F., Barchas, J. D., Clark, G. A., Donega, N., Kettner, R. E., Lavond, D. G., Madden, J., IV, Mauk, M. D., and McCormick, D. A. (1984). Neuronal substrates of associative learning in the mammalian brain. In *Primary Neural Substrates of Learning and Behavioral Change*, ed. by D. L. Alkon and J. Farley, pp. 71–100. Cambridge University Press.

Tsukahara, N. (1986). Cellular basis of classical conditioning mediated by the red nucleus in the cat. In *Neural Mechanisms of Conditioning*, ed. by D. L. Alkon and C. D. Woody, pp. 127–40. Plenum, New York.

Wigstrom, H., and Gustafsson, B. (1985). On long-lasting potentiation in the hippocampus: a proposed mechanism for its dependence on coincident pre- and postsynaptic activity. *Acta Physiol. Scand.* 123:519–22.

Woody, C. D. (1984). The electrical excitability of nerve cells as an index of learned behavior. In *Primary Neural Substrates of Learning and Behavioral Changes*, ed. by D. L. Alkon and J. Farley, pp. 101–28. Cambridge University Press.

16
Model building from molluscan and mammalian parallels

That *cellular* mechanisms of associative learning are conserved during evolution has now for the first time been given strong empirical support. The Ca^{2+}-dependent K^+ current intrinsic both to *Hermissenda* and rabbit neuronal membranes remains reduced for at least days after classical conditioning of the two animals. In a broad sense, this conditioning-specific change of membrane current is located in a cellular compartment common to both species. The reduction of Ca^{2+}-dependent K^+ current can be considered to be "postsynaptic," that is, it was not found in presynaptic terminal branches. In *Hermissenda* it was found in a cell body approximately 60 μm away from presynaptic endings. In the CA1 hippocampal pyramidal cell, it was recorded in the cell body, but the Ca^{2+}-dependent K^+ current reduction may have been located in dendrites, that is, postsynaptic sites of contact between the pyramidal cell and other neurons. The conditioning-specific reduction of the Ca^{2+}-dependent K^+ current could not have been located in CA1 presynaptic endings, because these are simply so far away that changes there would not have been observable with a microelectrode in the cell body. In any case, conditioning-specific changes of excitability were observable with microelectrodes in both *Hermissenda* and rabbit neuronal cell bodies. It should be emphasized that conditioning-specific membrane changes could *also* still occur in presynaptic endings – it is just that in no neurobiologic preparation has there been any clear evidence of such a location for storing associatively learned information. They have neither been ruled out nor clearly implicated as storage sites.

A postsynaptic locus and, more specifically, a dendritic locus for learning storage is most consistent with the design of the mammalian brain for maximizing information processing capability. As already mentioned, the vast number of convergences between inputs to neurons of the vertebrate central nervous system such as the CA1 cells are postsynaptic and involve dendritic loci that receive synaptic input from separate sources of presynaptic stimulation (such as a conditioned stimulus and an unconditioned stimulus).

The conditioned stimulus could cause, for instance, a postsynaptic potential change in a dendritic spine that is located near another dendritic spine in which a postsynaptic potential change is elicited by the unconditioned stimulus. With such proximity the potential change in one spine could spread to the other spine and thereby influence electrical and/or cellular events in a neighboring locus.

The interaction of visual and statocyst signals on the type B cell soma during *Hermissenda* conditioning provides a model for what could occur on the CA1 dendritic tree (Figure 66, Model Figure 1). The conditioned stimulus, light, causes an opening of Na^+ channels and a positive shift of membrane potential in the soma or cell body compartment of the cell. The unconditioned stimulus, rotation, causes synaptic effects on the terminal branches – a compartment at the other end of the type B cell, about 60 μm away. Rotation produces synaptic hyperpolarization (i.e., a negative shift of membrane potential) followed by synaptic depolarization. These synaptic shifts of membrane potential spread along the type B cell axon until they reach the cell body. The properties of the type B axonal membrane are such that as the synaptic potentials spread from the endings to the cell body, they only decrease by about 30%. When the conditioned and unconditioned stimuli are paired, the synaptic depolarization, but not the synaptic hyperpolarization, has an effect on the soma membrane. (This is because of differences in total membrane resistance as discussed in an earlier chapter.) Rotation-induced synaptic depolarization adds to light-induced depolarization. These potential changes activate the voltage dependent Ca^{2+} current and together with possible light-induced C-kinase activation thereby cause inactivation of K^+ currents.

The interaction of the visual stimulus with the rotation stimulus at least in part involves the electrical spread of a potential change in one compartment (for example, the synaptic endings) to another compartment (the cell body). (Of course, the converse could also easily occur, namely, light-induced depolarization could spread from the cell body to the synaptic endings. We do not know the nature of the synaptic ending membrane currents nor of their change with learning and thus, for our purposes, shall focus our attention on the type B cell body.) A similar type of interaction could occur between dendritic compartments of the CA1 pyramidal cells in the rabbit hippocampus. Synaptic shifts of membrane potential on one dendritic spine (due to an unconditioned stimulus) could spread to a neighboring spine that also receives synaptic input (due to the conditioned stimulus). With appropriate pairing, there could be summation of the two effects to activate (perhaps in a nonlinear manner) a voltage-dependent Ca^{2+} current, and, together with neurotransmitter-induced C-kinase activation, by the sequence previously described, cause a persistent change of postsynaptic excitability.

Postsynaptic signals at one dendritic locus, in a manner analogous to visual–statocyst interaction on the type B cell, could interact with signals at a second dendritic locus. These interactions provide the crucial elements for an entire conceptual framework or model for how a vertebrate brain could encode and store for later recall temporally associated stimuli. The interaction could begin electrically, as just described, and then be translated into differences of intracellular Ca^{2+} and kinase activation as observed for the *Hermissenda* type B cell. However, this need not be the only basis for interaction of neighboring dendritic loci. Intracellular messengers, possibly to a limited extent Ca^{2+} itself, may also diffuse between such loci. Intracellular spread of molecular messengers would, however, undoubtedly be a slower process than the electrical spread. There may, however, be a need for both electrical and slower intradendritic communication as an association between stimuli is learned.

Parallels with the cellular mechanisms obtained for *Hermissenda* conditioning can be considered together with the known cellular changes, behavioral characteristics, and psychophysical properties of mammalian associative learning, to constrain a vertebrate learning model. This would limit the possible models that might apply and thereby increase the probability for there being some physical reality for the model.

We know, for example, that behavioral and cellular responses to the unconditioned stimulus alone are not changed by conditioning, whereas responses to the conditioned stimulus are. In *Hermissenda*, because the conditioning-specific increase of membrane excitability is a particular locus, the soma or cell body, it will enhance cellular and behavioral responsivity to subsequent presentations of light but not to rotation. The postsynaptic response of the type B cell to subsequent presentations of rotation alone will not be significantly affected by a conditioning-induced change of soma membrane excitability. Furthermore, the postsynaptic response to rotation alone is sufficiently small so that the conditioning-induced reduction of K^+ currents in the soma will not become manifest. Only at the soma, when light causes a substantial voltage change (such as 25–35 mV), will the K^+ current reduction have an observable impact on the response characteristics (i.e., the manifest excitability of the type B cell). There must be, therefore, some asymmetry in the site for receiving input from the conditioned stimulus as compared to the site for receiving input from the unconditioned stimulus. This asymmetry may involve, as it does for the type B cell, large differences between the magnitude of voltage change elicited by the conditioned and unconditioned stimuli. It could also involve other differences such as those affecting the degree to which intracellular Ca^{2+} can rise, or the degree to which other biochemical consequences can occur.

In constructing our model of vertebrate associative learning, we will represent the necessary asymmetry as simply a difference in the magnitude of the voltage change elicited by the conditioned and unconditioned stimuli (just as with the light and rotation effects on the type B cell). The magnitude of the voltage change elicited by the conditioned stimulus will be greater than that elicited by the unconditioned stimulus but will become still greater as a result of stimulus pairing (Model Figure 1).

The cellular analysis of *Hermissenda* learning indicated that the genetically specified neural system (i.e., the organization of synaptic interactions) limits and defines the potential for associative learning. The necessary convergence points must first provide a pathway for two temporally associated stimuli to meet within the nervous system before a predictive relationship between the stimuli can be learned. This means that in the vertebrate nervous system, there must be a vast number of

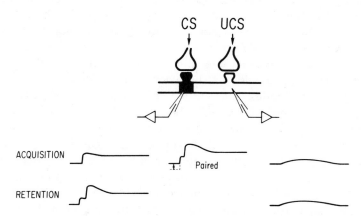

Model Figure 1. Voltage responses recorded from hypothetical dendritic sites postsynaptic to a conditioned stimulus (CS) and an unconditioned stimulus (UCS). During acquisition the CS-elicited voltage response is greater than the UCS-elicited response. During retention the CS-elicited voltage response remains enhanced and the UCS voltage response is unchanged.

convergences already existing within the nervous system to provide the potential for learning the vast number of associations within a mammalian capability. Thus, for any given UCS and any particular CS, there must be in our model, hosts of the dendritic elements just described. There also must be sufficient separation of these elements so that a learned relationship is specific and not confused with, or obscured by, other learned relationships. The dendritic tree of a neuron such as the CA1 cell provides both the large *number* of dendritic elements as well as the necessary separation of these elements. The postsynaptic site on element 1 for receiving the CS_1 input can, with conditioning, undergo Ca^{2+}-mediated K^+ current reduction and thereby increased excitability (Model Figure 2). Enhanced excitability at this site will only become manifest with sufficient voltage change during subsequent presentations of CS_1. After conditioning, CS_1 elicits a larger voltage change in the postsynaptic site and thereby more readily elicits impulse activity in a particular cortical cell (here "cortical cell I"). As a result of CS_1–UCS_1

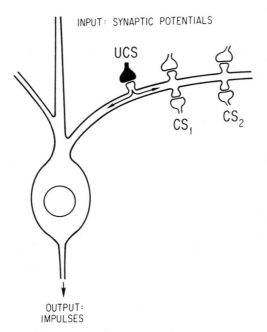

Model Figure 2. Pairing CS_1 and UCS synaptic potentials (input) modifies subsequent CS_1 synaptic potentials and, thereby, CS_1-elicited impulses (output).

pairing, however, none of the other inputs either within the same dendritic element (1) or within the same dendritic tree elicit a larger voltage change.

Now it might be asked if CS_1 can, as a result of conditioning, elicit a larger voltage change and thus more readily elicit a cortical cell I impulse, where is the specificity of encoding if this can occur following CS_1–UCS_1 pairing, or CS_1–UCS_2 pairing, or CS_1–UCS_3 pairing... etc.? Where is the predictive relationship if any one of many such possible pairings will enhance CS_1-elicited output? The predictive relationship can be preserved, the specificity maintained, if we now include not only a large number of dendritic elements, but also a large number of cortical cells with dendritic elements in our model. The neural representation of any given CS–UCS relationship will require an entire *set* of cortical cells to have dendritic elements altered by the conditioning (Model Figure 3). And although the set of such cells for one CS–UCS relationship will share cells with the set of cells for another CS–UCS relationship, the two sets will not be identical – there will be cells not common to both sets. The specificity of representation for any particular association will depend upon which *combination* of cells are more easily excited by subsequent presentations of a conditioned stimulus.

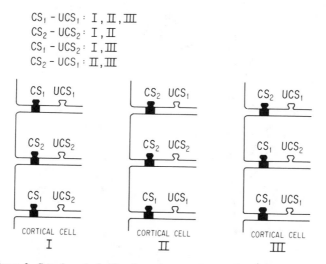

Model Figure 3. Set of cortical cells whose dendritic elements can be altered by conditioning. Combinations of I, II, and III determine specificity of CS–UCS learned associations.

This means that a surprisingly large percentage of cells within a defined population, such as the CA1 cells of the hippocampus, may participate in the storage and recall of an association without losing substantial storage capability for other associations. The model as constructed thus far is consistent with what was actually found in hippocampal slices from classically conditioned as compared to control rabbits. Amazingly, 50–60% of the CA1 neurons showed (by reduced amplitude and duration of the afterhyperpolarization) evidence of persistent $I_{Ca^{2+}-K^+}$ reduction. Parallel to these in vitro intrinsic changes of CA1 membrane currents, in vivo studies had previously revealed that 50–60% of CA1 cells responded after conditioning with more impulses in response to the conditioned stimulus. These results and the model derived from them (as well as the *Hermissenda* observations) recommend a dramatic departure from our notions as to how neural systems function in a mammalian brain. Excitation of a substantial fraction of neurons in any given population may be involved in learned responses to even very discrete and specific stimuli and patterns of stimuli.

Excitation of a fraction of a neuronal population may itself be considered as a pattern of neuronal responses. Obviously, the pattern elicited by one stimulus, after it has become associated with a second, unconditioned stimulus, must in some respects resemble the pattern elicited by the unconditioned stimulus. Otherwise, the conditioned stimulus will not become a behavioral predictor of the unconditioned stimulus and the conditioned response will not bear any similarity to the unconditioned response. Thus, there must be in our model some specificity in the pattern of cortical cells' firing for the genetically specified unconditioned stimulus effects. Specificity can be provided if we design the model so that any particular unconditioned stimulus can interact (i.e., converge) with a whole range of conditioned stimuli – but on the same cell – within the dendritic tree of one neuron. This is not to say that the unconditioned stimulus cannot effect more than one neuron – it almost certainly is presynaptic to an array of neurons. It is to say that when a conditioned stimulus becomes associated with the unconditioned stimulus (i.e., the CS predicts the UCS), it will elicit responses from the same set of neurons from which responses are elicited by the unconditioned stimulus. This would be true for as many conditioned stimuli that can become associated with a particular UCS. It should be possible that at least two different kinds of cortical arrays store learned associations. The first type would not be organized by the nature of the unconditioned

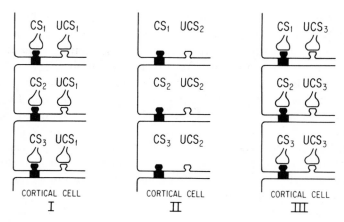

Model Figure 4. Organization by unconditioned stimuli (UCS) for a set of cortical cells whose dendritic elements can be altered by conditioning.

response (Model Figure 3), whereas the second type would have such an organization (Model Figure 4). The first cortical array type might, however, be organized in a different way such as by shared properties of CS inputs or their relevance to certain kinds of sensory experience. The first type of cortical array would be less directly responsible for actually eliciting a conditioned response. These latter arrays may be more typical of a structure such as the hippocampus.

Within the design of an array organized by unconditioned responses, another psychological phenomenon might be accounted for. Stimulus generalization is the facilitation of learning one CS–UCS relationship by the prior learning of another relationship with the same UCS but a different CS. If the same dendritic trees are activated by different conditioned stimuli, there may be some proximity of postsynaptic loci for receiving distinct presynaptic inputs. If one such locus becomes more excitable in response to CS_1, a nearby CS_2 may more readily excite the dendritic branch common to both CS_1 and CS_2 inputs and thereby more readily undergo change when paired with the unconditioned stimulus. This suggests some logic to the genetically programmed distribution of presynaptic inputs (potential conditioned stimuli) on a dendritic tree. Presynaptic inputs that are activated by environmental stimuli that have some resemblance to each other might be grouped together on the dendritic trees of neurons activated by unconditioned stimuli. This could be represented by Model Figure 2.

Interactions within and between cortical arrays

There are other interesting physiologic properties that must be sought for our model and that should have some correspondence to and be motivated by additional psychophysical phenomena. Often we remember many parts of a past experience or situation but not quite all the parts. We cannot quite recall some detail(s), which we know belong to that memory. We recognize a face, for example, but we cannot quite remember where we saw the face before – we cannot "place" it. We think for a moment about the face and suddenly we *associate* a name – Frank: Ah, yes, I met Frank in Paris five years ago. So first our memory is stimulated by the visual experience of the face, then we recall a name, and then, finally, we more closely recreate an entire experience, which includes a place, a time, and other circumstances (e.g., attending a meeting, etc.). It is as if a chain reaction of associations has occurred in which one association, or group of associations, triggers another, which in turn triggers another, etc. By the time this chain reaction has occurred, we are giving most, if not all, of our attention to it. To some extent, then, for full awareness and recognition of a past experience, we are involving a substantial portion of our brain. We are focusing on certain associations at the expense of others. This type of global process would almost certainly require a means of interaction between many of the associations involved in the recollection as well as many not involved. The chain reaction of one association triggering another could be a facilitatory interaction between associations, and the focusing, that is, the filtering out of other irrelevant associations as well as irrelevant sensory input, could be an inhibitory process. How might such features be incorporated into the design of our model?

First, it might be helpful to depart a bit from thinking only of conditioned and unconditioned responses. For such responses cortical arrays were organized by the need for expression of the learned association – the behavioral response to the unconditioned stimulus is now, as a result of conditioning, also elicited by the conditioned stimulus. A given cortical cell necessary for such a function may receive only one unconditioned stimulus but many conditioned stimuli. The chain reaction of remembering an entire experience may not require any behavioral response at all, but rather a large set of sensory images. We might think now of a different basis for organizing synaptic inputs to a dendritic tree. Instead of reference to the generation of a behavioral response, reference to the interaction, or lateral spread, between associations might be more useful. The distinction we are making in our model for the organization of synaptic inputs to a cortical dendritic tree might be described as depending on a sensory (input) rather than a motor (output) emphasis. With reference to conditioned responses, we emphasize how the *output* of the system might be reflected by the actual neuroanatomic organization of certain neural systems within the brain. With reference to a complex "recognition" experience, we emphasize how the *input* of the system might be reflected by neuroanatomic organization.

First, it is important to specify that we are now imagining the activity of vast numbers of neurons for a particular visual image such as that of a face. This vast number or "field" of neurons would consist of large numbers of neuronal sets similar to the cortical arrays discussed previously. Sets of neurons within the same vast array or field, or within different arrays could interact with each other. Inputs from distinct neuronal sets could converge on cortical neurons and interact via mechanisms defined for *Hermissenda* classical conditioning and paralleled in our model of vertebrate classical conditioning. Field inputs with sufficiently close postsynaptic location will interact with each other to induce a persistent change of excitability (Model Figure 5). Subsequent stimulus presentations that activate a neuronal field (i.e., a particular pattern of sensory stimuli) will – perhaps somewhat like a chain reaction – elicit activity in a series of neuronal sets. Instead of a discrete conditioned stimulus eliciting a discrete conditioned behavioral response, there would be a sensory constellation or pattern that would elicit a number of electrophysiologic responses from a constellation of neurons which would trigger responses from other constellations of neurons. The elicited electrophysiologic responses from a number of neuronal populations, then, together would generate a "picture" – would fill in the holes of a now recognized experience.

The convergence of many inputs from neuronal sets onto a dendritic tree need not be between only two distinct inputs (as was the case for a conditioned stimulus converging with an unconditioned stimulus). Instead, presynaptic endings from many different neurons may converge on the same dendritic tree. And whereas only a few such neurons may be insufficient to cause an excitability change, the collective effects of the group of inputs might be necessary to cause persistent excitability changes in the postsynaptic subfield neuron (Model Figure 5). Now the rules for the psychophysical properties, as well as the cellular mechanisms, might be somewhat different from the rules of conditioning. (However, the basic biophysical and biochemical sequence involving

Ca^{2+}-mediated reduction of K^+ currents for producing persistent excitability differences could be the same.) For instance, the temporal order of associations or associative elements might not be crucial, as is the case for classical conditioning. Stimulus inputs may only have to occur within a certain temporal interval of each other with no requirement for which stimulus is first, second, third, etc. (unlike the conditioning requirement that the conditioned stimulus precede the unconditioned stimulus). Similarly, when we remember elements of an experience, it is the total experience of temporal related events and stimuli, independent of their exact order, that we later recall. And our recall is not just in response to one of the stimuli (for instance, the conditioned stimulus) but in response to a variety of subsets of the original set of stimuli during the experience.

The role of *inhibitory* synaptic interactions in focusing or sharpening the recall of an image is suggested by their contribution to memory storage and recall in *Hermissenda*. As discussed earlier, the learned association of light with rotation is stored by an increased excitability of

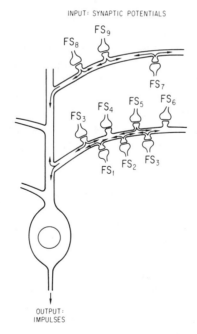

INPUT: SYNAPTIC POTENTIALS

OUTPUT:
IMPULSES

Model Figure 5. Postsynaptic sites for interaction of neuronal sets from "fields" ($F_1 \ldots F_n$) of neuronal inputs.

the type B cell (due to persistent Ca^{2+}-mediated K^+ current reduction) and decreased excitability of the type A cell. The two changes of excitability are complementary in reducing the synaptic excitation of a motor pathway responsible for controlling movement of the animal toward a light source. Conditioning has caused a new network bias in favor of the B cell and against the A cell. Even before conditioning, the type B cell caused synaptic inhibition of the type A cell. This synaptic inhibition will exaggerate differences of excitability brought about by conditioning. Because the type B cell is more excitable, it will cause more synaptic inhibition and thereby reduce the impulse activity output of the already less excitable type A cell. Similarly, we might include synaptic inhibition in our model design of cortical field interactions (and perhaps even of conditioned–unconditioned stimulus interactions). A particular image could activate an entire field of neurons, which in turn cause synaptic inhibition of nonactivated surrounding neurons (while activating other neurons). Such synaptic inhibition could also account for, or at least contribute to, the psychophysical phenomenon of "blocking." A learned association between two sensory stimuli, or constellations, blocks the learning of an association between one of these stimuli and a third novel stimulus. Synaptic inhibition due to the prior learning of an association could make it more difficult for a third set of neurons to become excited as a result of converging inputs. This inhibition may also explain the occasional experience of trying, but being unable, to remember something, this experience being rectified by rest and temporary cessation of all efforts and recollection. By removing the stimulation that might be "blocking" (or inhibiting) the recall of the memory of interest, the necessary neurons may be relieved from synaptic inhibition and thus become more accessible to later activation.

To summarize our hypothetical synthesis from *Hermissenda* and mammalian observations, we can consider that a pattern of stimuli within the environment activates a sensory field or vast array of neurons. Activation of this field based on past learning experience will, in turn, preferentially activate other sets of neurons that receive many converging synaptic inputs from the initially stimulated neuronal set or field. When these neuronal sets – representing associated elements in a recognized or recalled experience – are sufficiently activated, they cause inhibition of most other neurons. This synaptic inhibition could both focus attention on the activated neuronal subfields and "sharpen" the remembered image. This process of sensory field activation, secondary activation of other neuronal sets, inhibition, focusing, and sharpening may provide the basis for involvement of "the whole brain" in remembering, that is, how an image is ultimately recognized and how we pay attention to it.

Bibliography

Alkon, D. L. (1984). Calcium-mediated reduction of ionic currents: a biophysical memory trace. *Science* 266:1037–45.
    (1987). Conservation of cellular mechanisms for models of learning and memory. In *Cellular Mechanisms of Conditioning and Behavioral Plasticity*, ed. by D. L. Alkon, W. Crill, J. McGaugh, P. Schwartzkroin, and C. D. Woody, Plenum, New York. In press.
Konishi, M. (1986). Centrally synthesized maps at sensory space. *TINS* 9:163–8.
Mishkin, M., Ungerleider, L. G., and Macko, K. A. (1983). Object vision and spatial vision: two cortical pathways. *TINS* 10:414–17.
Mountcastle, V. B., Lynch, J. C., Georgopoulos, A., Sakata, H., and Acuna, C. (1975). Posterior parietal association cortex of the monkey: command functions for operations within extrapersonal space. *J. Neurophysiol.* 38:871–908.
Pellionisz, A. (1970). Computer simulation of the pattern transfer of large cerebellar neuronal fields. *Acta Biochim. Biophys. Acad. Sci. Hung.* 5:71–9.
Szentágothai, J., and Arbib, M. A. (1975). Conceptual models at neural organization. M.I.T. Press, Cambridge, Mass.
Wiesel, T. N., Hubel, D. H., and Lam, D. M. K. (1974). Autoradiographic demonstration of ocular dominance columns in the monkey striate cortex by means of transneuronal transport. *Brain Res.* 79:273–79.

17
Mollusc versus mammal: nonparallels

The similarity of the K^+ current reduction intrinsic to neuronal membranes of classically conditioned *Hermissenda* and classically conditioned rabbits is striking. The implication of Ca^{2+}- and lipid-dependent phosphorylation in the regulation of K^+ currents during and after conditioning of both mollusc and mammal is also clear. The possibility of cellular mechanisms of learning and memory common to such diverse species is therefore plausible. But as discussed earlier, the very nature of the neural systems in these animals almost dictates phenomenologic differences in both the behavior and electrophysiology of learning. The spatial and temporal resolution cannot be the same for a system with billions of neurons as opposed to thousands, with action potentials lasting a few milliseconds as opposed to tens of milliseconds, with synaptic potentials occurring with a delay of a fraction of a millisecond as opposed to many milliseconds. The capacity to process and integrate information is vastly greater in the mammalian brain.

We may construct elementary conditioning building blocks from the same ionic channels, the same Ca^{2+}-mediated biochemical regulation of these channels, and the same convergence of distinct sensory inputs onto postsynaptic structures. But the capability of a highly developed brain permits functions simply absent from a snail's repertoire. Some functional differences are familiar to us all. They involve the capacity for abstract thinking, for language, etc. One, of which we frequently are not aware, involves to some extent an undermining or control of the memory process itself.

The adaptive value of learning, or predicting stimulus relationships in the environment, is obvious. Less obvious is the value of forgetting–not forgetting, simply through a lack of frequency of encountering what was remembered, or forgetting due to *un*learning, that is, encountering stimuli previously related but now unrelated to each other (as with extinction described earlier). There is another kind of forgetting that involves a "walling off" – removing a memory from our conscious awareness. The memory itself is actually preserved, but our attention to it – our ability even to recall it if we wish is gone, if not greatly reduced. The adaptive value of this type of forgetting is better understood in the context of our own broad design as organisms.

The capacity to sense, learn, and remember must always be considered in the service of satisfying our basic needs: the recurrent reduction of drives that defines our very existence as purposeful beings. We are organisms always in a state of unstable equilibrium, an equilibrium regularly thrown out of balance by hunger, fear, pain, etc. Our memories not only serve to motivate behavior that will ultimately reduce or satisfy such drives, but also, themselves, bring some such satisfaction, or the lack thereof. The very anticipation of future experience or reflection on past experience can provide a source of pleasure or of pain. Pleasurable experience, of course, heightens our attention, increases our perception, and thereby aids our memory. The degree of attention will control the subsequent emotional content of a memory-elicited response. Attention opens the gate for recruiting the appropriate emotional force. Painful experience cannot only cause us to shift our attention, to filter our perception; it can also cause us to forget what has actually occurred. The *biological record* of the painful memory is still there – yet it is forgotten. How such forgetting is actually accomplished physiologically, of course, is a matter for speculation. It does seem to require, however, a willful remembering of other less painful experiences, and as such would involve a very sophisticated means for reducing painful associations. It appears that by learning to recall these other memories we divert our attention from unwanted memories that are actually present in our biological archives. This might be considered a form of "blocking," as discussed earlier, but is probably achieved by quite different mechanisms. That such forgetting – or burying of painful memories – does occur highlights the presence of at least two separate stages in our own recall process. The first stage is the activation of neurons that, in the right combination, are actually storing the memory. The second stage is paying attention to, becoming aware of, becoming, as it were, conscious of the memory.

18
Biological records of memory

We began by asking: What are the biological means for actually recording a learned experience? What is it that gives memory physical reality? Thus far in both *Hermissenda* and rabbit neural systems, we have found a biophysical memory trace. Reduction of the same K^+ currents in molluscan and mammalian brains provided a means of altering the excitability of neurons at least for many days. In *Hermissenda* it was possible to reconstruct a causal sequence that could account for how such K^+ current changes arise during learning and how they endow relevant neural networks with new properties that are the cellular translation of behavioral features characteristic of classical conditioning.

There is considerable evidence for the involvement of biochemical steps in the production of this biophysical memory trace. Calcium- and lipid-dependent phosphorylation of low molecular weight proteins appears to facilitate the transition, during conditioning, from biophysical phenomena that last for fractions of a second or seconds to phenomena lasting for days or longer. "Or longer," of course, defines another frontier. Are the K^+ current changes, themselves, sufficiently permanent to account for memories that last for a lifetime? Or, do they accompany or help generate other neuronal transformations responsible for permanent memory storage? For *Hermissenda*, conditioning-specific alteration of m-RNA turnover reached its maximum about one day after training, and although this increased turnover became much less marked it still persisted four days after training.

Thus far, in other invertebrate learning preparations, such as the octopus, the bee, the locust, and other gastropod molluscs (e.g., *Aplysia*), we have not yet uncovered biological records of associative memory. Many interesting hypotheses have been formulated. Electrophysiologic differences have been observed for many minutes and even hours. But no persistent biophysical, biochemical, or structural alterations have been

158

shown to be specific to the acquisition and retention of a learned association by these animals.

Vertebrate preparations, however, have yielded some interesting alternatives to the memory traces found in *Hermissenda* and now the rabbit hippocampus. Prolonged exposure of animals to marked alterations of sensory experience apparently can influence the complexity and nature of dendritic branching patterns of cortical neurons. Animals maintained in "rich" environments (i.e., with abundant sensory stimulation) have neurons with greater numbers of dendritic spines, postsynaptic loci. Animals maintained with restricted or deprived sensory stimulation have brain regions that show gross derangements in subsequent responsiveness to physiologic stimuli as well as obvious morphologic abnormalities. These are clearly structural changes in neural systems that only result from long periods (weeks or more) of environmental manipulation. They typically occur most readily (and sometimes exclusively) during critical periods of an animal's development. What cannot yet be determined is the relation of such structural changes, arising from gross environmental manipulation, to the physiologic storage of learned stimulus relationships. It may be that these structural changes define an endpoint that eventually results from the types of biophysical and molecular modifications observed in *Hermissenda* and the hippocampal slice. Alternatively, drastic structural changes may be in a class of biological phenomena quite distinct from those that underlie learning. There is no doubt that brain alteration with sensory deprivation or enrichment, particularly during critical development periods, will influence what can or cannot be learned (as well as sensed). It remains an open question, however, whether the learning process itself requires expression in the actual shape and geometry of neurons.

As with all aspects of learning physiology, including the involvement of vast arrays of neurons in pattern recognition, storage, and recall, we will have to develop techniques to study the thing itself. We will have to measure in complex brains the relevant parameters during and after the acquisition of discrete stimulus associations. Learning-induced structural changes, distributions of learning-induced modifications in entire neural networks, etc., might be revealed by exploiting insights (only recently achieved) into membrane and molecular mechanisms of memory storage. For example, immunologic recognition of the C-kinase in cytoplasmic and membraneous compartments with a specific antibody might provide a means of tracing through a nervous system, simple or complex, a new path of information flow during the recall of a learned association. Ultimately, a biological memory record may, when suitably analyzed, assume the form of an image or pattern that is recognizably unique and subtle. For the present we have no such image, but rather what seem to be a few of its crucial elements. What in *Hermissenda* may comprise much, if not all, of a memory record, in the rabbit hippocampus or cerebellum represents evidence of conserved mechanisms for generating a part of such a record, rather than the complete record itself. We are, nevertheless, closer to understanding what it is we are looking for and how (conceptually and experimentally) we should look.

Bibliography

Greenough, W. T., Juraska, J. M., and Volkmar, F. R. (1979). Maze training effects on dendritic branching in occipital cortex of adult rats. *Behav. Neural Biol.* 26:287–97.

Horn, G., Rose, S. P. R., and Bateson, P. P. G. (1973). Experience and plasticity in the central nervous system. *Science* 181:506–14.

Hubel, D. H., and Wiesel, T. N. (1965). Binocular interaction in striate cortex of kittens reared with artificial squint. *J. Neurophysiol.* 28:1041–59.

Appendix
Neural systems at *Hermissenda*

Visual system

Photoreceptors – sensory or primary cells

In each of the two *Hermissenda* eyes there are five photoreceptors: Two called type A and three called type B. Each of these cells can be unequivocally distinguished from each other based on their locations, their size and shapes (Figure 21), their electrophysiologic properties (Figures 78–81), and their synaptic interactions (Figures 21 and 82).[1] The *Hermissenda* photoreceptors are specialized to convert or transduce light energy into electrical signals such as action potentials. The photoreceptors, in addition, have membrane properties and make synaptic interactions that serve integrative functions usually reserved for more central neurons in the vertebrate brain. This "peripheralization" of neural function is typical of less evolved species and exemplifies a comparative lack of specialization in neural function. In molluscs such as *Hermissenda*, a given neuron such as a photoreceptor may have to

---

[1] The type A cells are located near the lens and ventrally. They have spike amplitudes of approximately 45 mV (Figure 55), are not spontaneously active in darkness, and are not sensitive to dim light (Figure 56). Impulse trains of type A cells do not show sustained high frequencies of impulses (i.e., they accommodate). Type B photoreceptors have spike amplitudes of approximately 15 mV, are spontaneously active in darkness, and are sensitive to dim light (Figures 55–57). Type B impulse trains show little accommodation. Impulse trains of type A cells show clear accommodation. Other characteristics help differentiate the five photoreceptors from each other still further. The medial type A photoreceptors, for example, show small discrete depolarizing waves, whereas one of the type B receptors, the most sensitive to dim light, tends to have impulses that occur in couplets following bright lights. Type B photoreceptors are somewhat elliptical in shape with a long axis of 30–40 μm and a short axis of 15–25 μm. Type A cells are more spherical with a diameter of about 20 μm. (In response to light, all five of the photoreceptors depolarize and show greatly increased impulse activity.) On the ultrastructural level, the photoreceptors show extensively infolded membranes on which are located rhodopsin molecules that help transduce photic energy into electrical signals via a photoisomerization. The cell bodies of the photoreceptors are filled with small clear vesicles that are believed to contain acetylcholine.

161

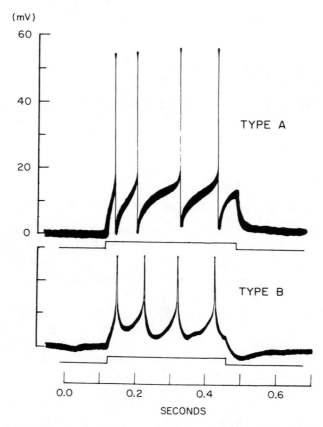

Figure 78. Intracellular recordings of type A and type B photoreceptor impulses elicited by positive current (0.4 nA, upper; 0.3 nA, lower) injection.

perform several neural functions (see below) that might be provided for with different individual neurons in higher nervous systems with a vastly greater number of neurons available.

The photoreceptor depolarizes (Figures 79–82) in response to light, that is, the voltage difference across its membrane becomes more positive. This light-induced shift of membrane potential spreads from the cell body (where transduction occurs) down the axon to a particular region of membrane where action potentials are triggered (Figures 21 and 83). These action potentials then propagate further along the axon into a spray of fine branches where neurochemicals or neurotransmitters are released onto the fine branches of synaptically linked neurons. (When

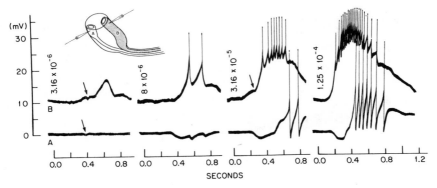

Figure 79. Responses of A and B cells to flashes of light. A dim flash (3.16 × 10⁻⁶ ergs/cm²-sec) produces a noisy depolarization in the B cell (upper trace) with no effect on the A cell. Note the small hyperpolarizing potential (arrows) appearing simultaneously in the two cells. With a flash of intensity (8 × 10⁻⁶), the B cell produces two spikes and the A cell develops a hyperpolarizing wave with two superposed wavelets. The onset of the hyperpolarization of the A cell precedes the depolarization of the B cell. Bright flashes evoke firing of the A cell, associated with interruption or slowdown of the discharge of the B cell. (From Alkon and Fuortes, 1972)

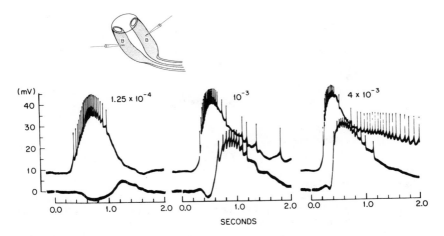

Figure 80. Responses of B cells to flashes of light. The more sensitive cell in the upper trace responds to a flash of intensity (1.25 × 10⁻⁴ ergs/cm²-sec) with large generator potential, whereas the cell in the lower trace develops a hyperpolarizing wave that delays the visible onset of the generator potential. With brighter flashes the generator potential in the lower cell starts early but is interrupted by a hyperpolarizing wave that becomes smaller as flash intensity increases. The firing of the more sensitive cell is decreased in coincidence with the peak of the response of the lower cell. (From Alkon and Fuortes, 1972)

163

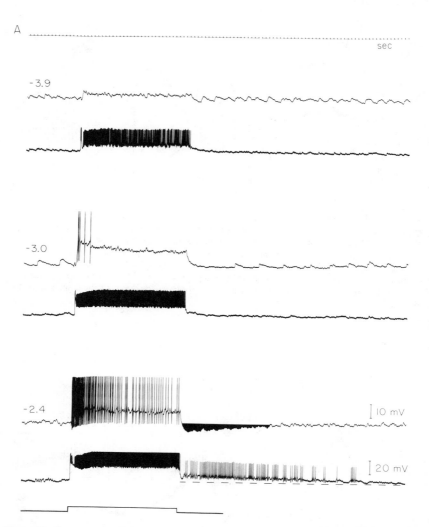

Figure 81. Responses to light steps of simultaneously impaled photoreceptors. Intact type A and type B photoreceptors are exposed to steps (indicated by the bottom trace) of increasing (top to bottom) intensity (expressed in − log units on the left of the traces). With sufficient light intensity an LLD (a long-lasting depolarization following the cessation of light) appears in the response of a type B photoreceptor (the lower trace in each pair of records). For this same intensity and with the same time course, an LLH (a long-lasting hyperpolarization following the cessation of light) appears in the response of a simultaneously impaled type A cell. For illustration purposes the difference of the type A membrane potential from the resting level during the LLH is represented by the darkened area. (From Alkon and Grossman, 1978)

164

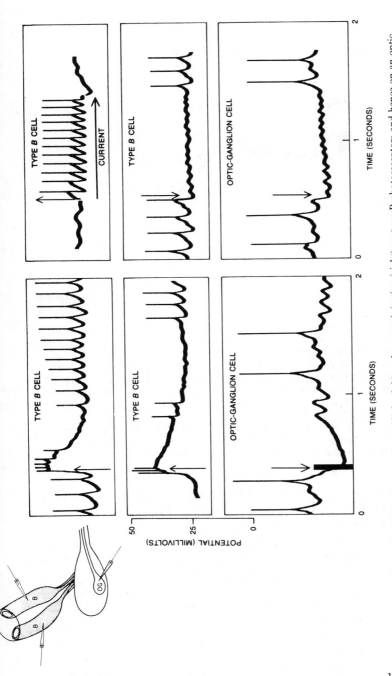

Figure. 82. Intracellular recordings show the effect of light (left) and current injection (right) on type B photoreceptors and hence on an optic ganglion cell they inhibit. In response to a brief flash of light (bar), two B cells become depolarized, or more positively charged (upward pointing arrows). These signals generate inhibitory postsynaptic potentials in the optic ganglion cell, making it more negatively charged (downward pointing arrow); fewer impulses are triggered. Injection of a positive current (horizontal arrow) into either of two B cells simulates the effect of light, eliciting impulses that inhibit both the other B cell and the optic ganglion cell. (From Alkon, 1983)

165

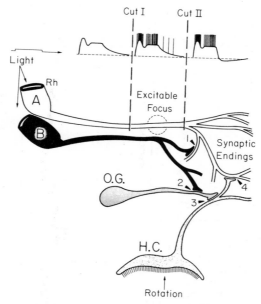

Figure 83. Excitation and inhibition of a type A photoreceptor. A type A photoreceptor soma with rhabdome, axon, excitable focus, and terminal branches is diagrammed. Light depolarizes photoreceptors at rhabdomes (Rh). Type B photoreceptors (of which there are three) inhibit, at synaptic endings, lateral type A, directly (1) and indirectly (2) by inhibiting optic ganglion (O.G.) cells (of which there are 13). Ipsilateral hair cells (H.C.) receive less inhibition from optic ganglion cells (3) when the type B fires and thus increases its inhibition of the lateral type A (4). Hair cells also inhibit type A cells when they depolarize in response to rotation. In response to light, the type A cell depolarizes without impulses or afterhyperpolarization with a cut I lesion. It depolarizes with impulses but without afterhyperpolarization with a cut II lesion. The response of an intact type A cell is represented at the upper right. (From Alkon and Grossman, 1978)

the type B cell's action potentials release neurotransmitters onto another cell, the type B cell is called "presynaptic," or before the synapse, and the receiving cell is called "postsynaptic," or after the synapse.)

In addition to the structural and electrophysiologic features that make it possible to identify each of the photoreceptors, there are rules of synaptic interaction that have been shown by repeated observations to be obeyed from one adult animal to the next. Type A photoreceptors, for example, have no synaptic contact, whereas type B photoreceptors (Figures 21 and 82) are all mutually inhibitory, that is, action potentials of each of a type B cell cause synaptic effects that reduce the frequency of action potentials of the other type B cells. Action potentials of the medial type A photoreceptor cause an excitatory synatpic effect on

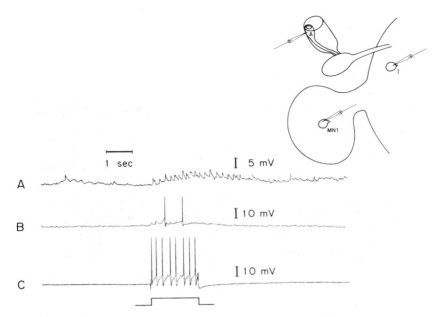

Figure 84. An example of triple simultaneous recordings from the medial A photoreceptor (C), the interneuron (B), and the MN1 cell (A). The interneuron and the MN1 cell were hyperpolarized by negative current injections of − 0.2 and − 1.0 nA, respectively. When the A photoreceptor was excited by positive current injection (+ 0.5 nA), the slightly hyperpolarized interneuron responded with EPSPs and two spikes, but the hyperpolarized MN1 cell showed many more than two EPSPs with somewhat prolonged EPSP occurrence. (From Goh and Alkon, 1984)

particular neurons, called "interneurons" (Figure 84), in the central ganglion, whereas the type B action potentials cause inhibitory synaptic effects on other cells within a small discrete aggregate called the "optic ganglion" (Figures 82 and 85).

Second-order neurons:
Optic ganglion cells and other interneurons

A number of different types of *Hermissenda* neurons receive synaptic input from the photoreceptors of the *Hermissenda* eye, that is, they are second-order stations for visual information as it proceeds through the nervous system. Cells of each of the two optic ganglia (each comprised of 14 neurons) receive inhibitory synaptic potentials produced by presynaptic action potentials of the type B photoreceptors (Figure 85). At least one of the optic ganglion cells, the S cell, controls excitatory synaptic effects on the type B cells (Figure 86).

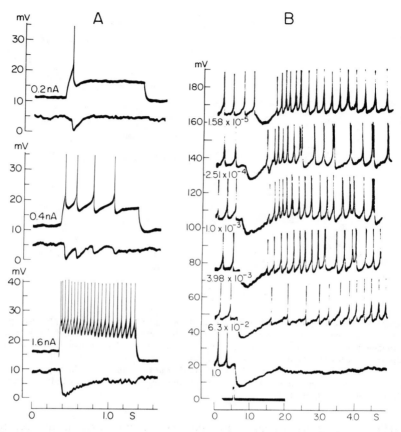

Figure 85. (A) Depolarizing currents to a type B photoreceptor produce impulses associated with IPSPs in a simultaneously impaled optic ganglion cell. (B) Responses of the C cell to light flashes of increasing intensity. Note the increased frequency of firing after the hyperpolarization for dim to moderate flashes. The duration of hyperpolarization with cessation of firing increases only for the brighter flashes. The values beneath intracellular voltage recordings refer to attenuation with neutral density filters of a quartz–iodide source with intensity on preparation of 10^5 ergs/cm^2 s designated as (1). (From Alkon, 1973)

Central neurons in the large neuronal aggregates, called ganglia, also receive synaptic input, in some cases inhibitory, in other excitatory, from type B photoreceptors (Figure 87). As already mentioned (Figure 84), one particular ganglial neuron receives excitatory synaptic potentials from the medial type A photoreceptor. Still other neurons, which receive synaptic input from the photoreceptors, are the sensory cells of the *Hermissenda* vestibular organ known as the statocyst.

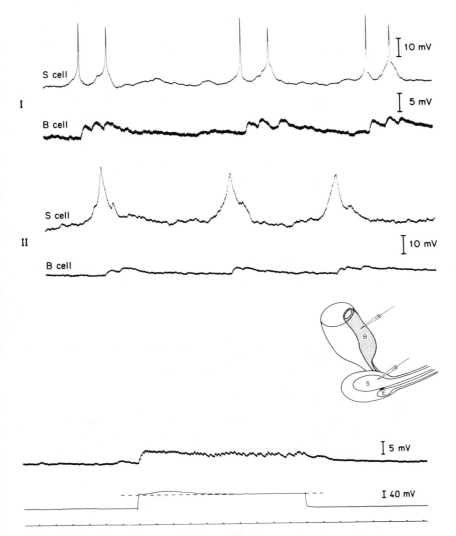

Figure 86. (A) Relation of S cell impulses to EPSPs recorded from type B photoreceptors. EPSPs were always preceded by S cell depolarizing waves. S cell impulses (inactivated in II after many minutes of recording) were not necessary to elicit EPSPs. Time scale: 1 s. (From Tabata and Alkon, 1982) (B) Effect of changes of S cell membrane potential on EPSPs recorded from type B photoreceptors. When the S cell (lower record) is hyperpolarized by a negative dc current (-1.6 nA), the EPSPs of a type B photoreceptor (upper record) are not present. Without current, the membrane potential of the S cell returns to its resting level and EPSPs appear. The type B cell was hyperpolarized by a negative dc current (-0.8 nA) to eliminate impulse activity during the experiment. Current injections into a type B cell did not affect the S cell. The dotted line indicates resting membrane potential. Time scale: 1 s. (From Tabata and Alkon, 1982)

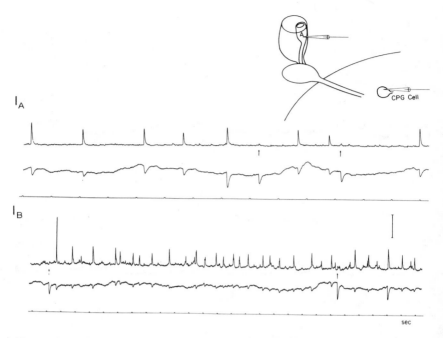

Figure 87. (I) Simultaneously evoked EPSPs in a second-order (CPG) visual neuron and IPSPs in a type A photoreceptor. *Upper trace*: CPG neuron. *Lower trace*: type A photoreceptor. The CPG neuron was hyperpolarized by injection of current (-1.0 nA). (II) EPSPs recorded from a second-order visual neuron associated with each impulse of a type B photoreceptor. *Upper trace*: The CPG neuron; this neuron was hyperpolarized at -40 mV from the resting level by injection of current (-2.1 nA). *Lower traces*: recording from a type B photoreceptor. (A) Responses to light as indicated by horizontal bar (intensity: -3.0). (B) Responses to depolarizing current step (0.22 nA) applied to the type B cell, as indicated in the lowest trace. The arrows indicate IPSPs in the type B cell and corresponding small EPSPs in the CPG neuron. Voltage calibration: 5 mV for upper traces, 10 mV for lower traces. (From Akaike and Alkon, 1980)

Motorneurons

The neurons of the "pedal" ganglia in a number of molluscs have been implicated as centers of motor control In *Hermissenda* pedal ganglia, several larger sized neurons that receive visual input have morphologic and physiologic features consistent with having roles as motorneurons, that is, they send axons that course through nerves exiting from the ganglia to the peripheral musculature, and their impulses can cause stereotypic movements of the animal. The largest neuron (pedal 1) in the entire nervous system, for example, controls movement of feather-like appendages known as cerata (Figure 32). Impulse trains of another

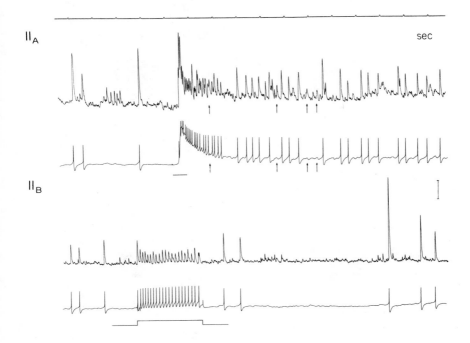

pedal neuron, MN1 (Figure 29), cause a turning movement of the animal. Since the frequency of MN1 impulses is increased by synaptic input from the medial type A cell via the aforementioned interneuron (Figure 84), the MN1 cell can participate in the turning movement of *Hermissenda* toward areas of maximal light intensity.

Ipsilateral–contralateral relations of the visual pathway

Thus far, synaptic relations of visual system neurons on only one side of the nervous system have been described. Photoreceptors of the right and left eyes do not interact directly but optic ganglion cell interactions are eminently suited to enhance differences of transmitted information regarding the intensity of illumination received by the two eyes. The axon of one type of optic ganglion cell, the C cell, gives off terminal branches that receive synaptic input from type B photoreceptors of the left and right eyes (Figure 27) on both the left and right sides of the circumesophageal nervous system. Impulses of the C cell cause inhibitory postsynaptic potentials on the terminal branches of con-

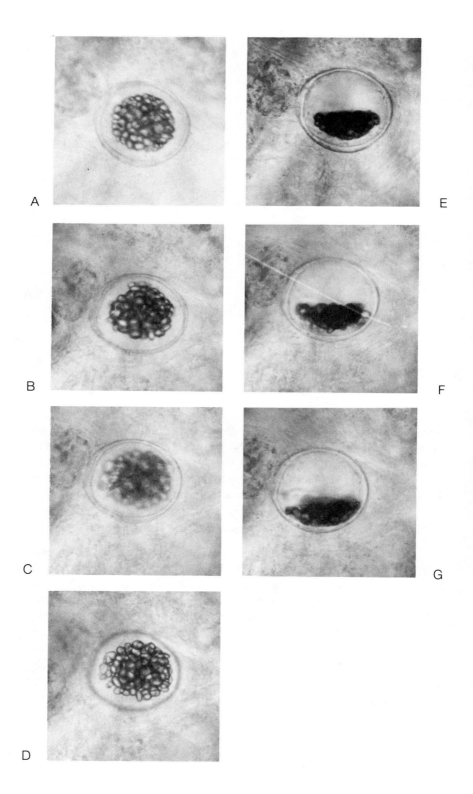

tralateral (i.e., on the opposite of the *Hermissenda* central nervous system) optic ganglion cells, the D cells whose axons extend only ipsilaterally (i.e., on the same side of the nervous system). Illumination, therefore, of left type B photoreceptors will cause increased numbers of type B impulses and thus increased numbers of IPSPs received by left optic ganglion cells (Figure 85), both C and D types. Left optic ganglion cell impulses, which occur with considerable frequency in darkness, will cease or occur with a much lower frequency in response to ipsilateral illumination. Right D optic ganglion cell impulses, however, will occur with increased frequency because in response to contralateral illumination they are much less inhibited by the left optic ganglic C cell impulses (which have been inhibited by the impulse activity of the left type B cells). The difference in impulse activity of right and left D cells therefore reflects, as amplified by the organization of the synaptic interactions, the difference of illumination received by the two eyes (Figure 28).

Vestibular system of Hermissenda

The statocyst is a transparent spherical structure (Figures 35 and 88) comprised mainly of the cell bodies of sensory receptors known as hair cells as well as supporting cells. There are 13 hair cells, which are concave discs 5–10 μm thick and ~ 50 μm in diameter. Cilia or "hairs" (Figures 35 and 89), 15 μm long, project from the inner surface of the hair cells into the statocyst lumen, which is filled with a fluid known as statolymph. The movements of these sensory cilia maintain 150–200 $CaCO_3$ crystals (known as statoconia) in a spherical cluster roughly centered within the lumen of the statocyst (Figure 88). Gravity tends to cause the crystals to clump in one half of the lumen and weight the cilia of specific hair cells (Figure 88). Hair cells with loaded hairs (i.e., weighted with the crystals) respond with depolarization (i.e., a positive change in the potential difference across the cell membrane) and increased frequency of action potentials (Figures 90 and 91). Rotation,

Figure 88. The effect of chloral hydrate (15 mg/ml) on statoconia distribution. (A)–(D) Viewed dorsally, (E)–(G) same preparation, viewed after tilting the microscope 90°. Note that, with longer exposure to chloral hydrate, the statoconia are more dispersed and out of focus (as viewed dorsally) because they fall toward the bottom of the statocyst, and touch the hair cells' inner surface (as seen by tilting the microscope). This indicates progressive collapse of the hairs. No statoconia or hair movement are observed for panels (C), (D), and (G). (A) and (E) Control; (B) and (F) 2.5 min of treatment; (C), (D), and (G) 5 min of treatment. The depth of focus increased from (C) to (D). (From Grossman et al., 1979)

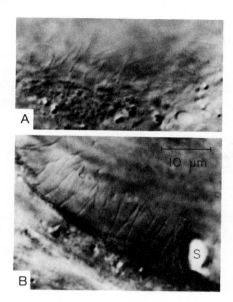

Figure 89. Visualization of motile hairs in the *Hermissenda* statocyst. Hairs were visualized with Nomarski optics (×1250) in an intact statocyst placed on a microscope tilted 90°. (A) Vital preparation. Apical half of the motile hairs could not be photographed but could be visualized. (B) Preparation fixed with 0.5% glutaraldehyde. The full lengths of the hairs, now immobilized, are apparent in the photograph. A single statoconium (S) is seen among the hairs. (From Grossman et al., 1979)

another stimulus that moves the crystals against the hairs of cells on one side of the cyst, produces similar voltage responses. Hair cells in front of the centrifugal force vector produced by rotation (i.e., hair cells whose hairs are compressed by the force of accelerated crystals) respond with depolarization and increased frequency of action potentials. Hair cells behind the force vector (i.e., whose hairs are not in contact with the statoconia) respond with hyperpolarization and decreased impulse activity.

Synaptic interactions between hair cells

Hair cells, within the same statocyst, that are located 180° apart (along an equatorial section through the statocyst) are mutually inhibitory (Figure 92). Hair cells 90° apart have unidirectional inhibitory interactions. Hair cells in the right statocyst can have inhibitory synaptic interactions or excitatory interactions with hair cells of the left statocyst.

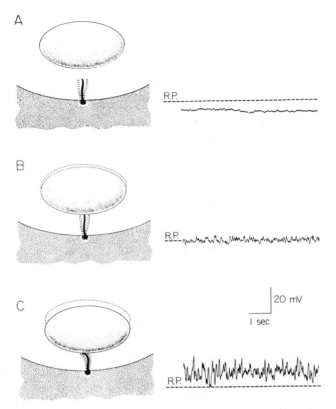

Figure 90. Conceptual model for hair cell voltage fluctuations (called "noise") and generator potential production. The hair is inherently motile (range of motion indicated by the dashed lines). (A) When the hair cell (oriented in the same direction as the centrifugal force vector) experiences a negative centrifugal force, the hair moves freely through a maximum range of motion producing little voltage noise (see noise record on right). (B) When the hair cell on that statocyst equator visualized when looking dorsoventrally is exposed to zero centrifugal force, the hair has some interaction with the statoconium. The statoconium exerts little net force on the hair although collisions of the statoconium with the rigid hair increase the voltage noise variance and frequency (voltage noise record on right). (C) When the hair cell (with luminal surface opposite the force vector) is exposed to a substantial centrifugal force (e.g., 1.0 g), the statoconium exerts a considerable net force on the hair, limiting its range of movement and reducing its movement frequency as well as dramatically increasing the voltage noise variance (see noise record on right). The increase of voltage noise variance and frequency results, by summation of individual events, in a depolarizing generator potential. The dashed line indicates the level of resting membrane potential (RP) before a negative centrifugal force (A) or a positive centrifugal force is exerted by rotation. All noise records were taken from actual hair cells under the conditions described. (From Grossman et al., 1979)

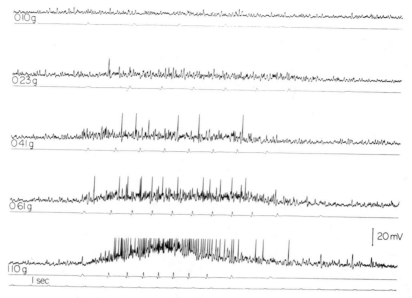

Figure 91. Response to rotation of the hair cell with luminal surface opposite to the centrifugal force vector. The axon of the hair cell was cut to eliminate all synaptic but not impulse activity. Maximal centrifugal force is indicated at the left of each lower trace monitoring the rotation. The amplitude of the monitor signal is proportional to the angular velocity of the turntable. The interval between monitor signals equals the period of the turntable's rotation. As centrifugal force causes progressively more weighting of the hairs, the voltage noise and steady depolarization of the hair cell increases and thus causes progressively more impulses. Impulse peaks are not included in the bottom record. (From Grossman et al., 1979)

Hair cell effects on central neurons

Hair cells cause EPSPs on the same interneuron (in the cerebropleural ganglion) that receive synaptic excitation from type B photoreceptors. Hair cells then also cause excitation of the MN1 cell in the pedal ganglion via this interneuron.

Interaction of Hermissenda visual and statocyst pathways

Receptor–receptor interactions

Hair cells with specific locations within the statocyst have specific synaptic interactions with the photoreceptors of the ipsilateral and contralateral eyes. Impulses of the caudal hair cell (i.e., the cell located toward the tail half of the cyst on the dorso–ventral equator) cause a cessation of EPSPs received by ipsilateral type B cells (Figure 93). This

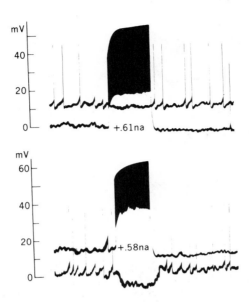

Figure 92. Reciprocal inhibition between hair cells in the same statocyst. Numbers refer to the strength of the depolarizing current pulses. In each frame the passive cell was slightly depolarized by passage of a steady positive current (0.10 nA, upper frame; 0.16 nA, lower frame). (From Detwiler and Alkon, 1973)

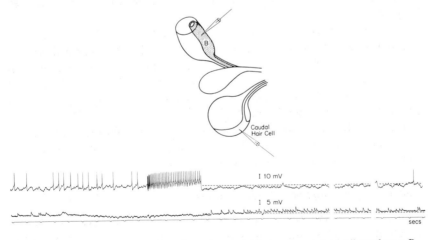

Figure 93. Simultaneous recordings from the caudal hair cell and the ipsilateral type B photoreceptor. The hair cell (upper recording) impulse train (elicited by a + 0.6-nA current step) caused little or no hyperpolarization but is followed by a long-lasting increase of EPSPs recorded from the type B cell. Again, the hair cell remains hyperpolarized following the impulse train, at least partially due to an increased frequency of IPSPs. The dashed lines indicate levels of resting potential for the two cells. Record interruptions are ~ 25 s each. (From Tabata and Alkon, 1982)

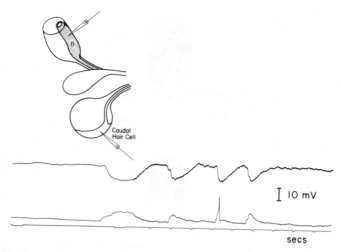

Figure 94. Simultaneous recordings from the caudal hair cell and the ipsilateral type B photoreceptor. Note that IPSPs recorded from the hair cell occur in synchrony with EPSPs recorded from the type B photoreceptor lower records. (From Tabata and Alkon, 1982)

effect is believed to result when hair cell impulses inhibit the E optic ganglion cell, which is the presynaptic source of the EPSPs received by the type B cell as well as IPSPs received by the caudal hair cell (Figure 94).

Another hair cell, adjacent to the caudal hair cell, is located at approximately 45° with respect to the caudal axis. This hair cell, known as the In hair cell, causes a synaptic inhibition of ipsilateral type B photoreceptors as well as a cessation of the EPSPs just mentioned (Figure 95). The In hair cell also causes synaptic inhibition of at least one of the two ipsilateral type A photoreceptors.

Type B photoreceptor impulses cause synaptic inhibition of the cephalic hair cells (Figure 96). They also cause cessation of large spontaneous IPSPs received by the caudal hair cell.[2]

Hair cell optic ganglion cell interactions

Caudal and In hair cell impulses trains cause inhibition of the E optic ganglion cell. This inhibition lasts throughout the impulse train and is

---

[2] Other receptor interactions, not mentioned thus far, include a nonsynaptic excitation (i.e., via accumulation of potassium ions in the extracellular space common to two or more neurons), of a single hair cell (believed to be the most ventrally located) by an ipsilateral type A photoreceptor. Inhibitory synaptic interactions also exist between photoreceptors and contralateral hair cells.

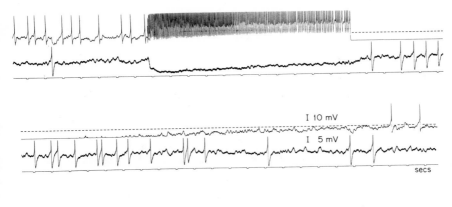

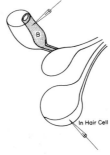

Figure 95. Simultaneous recordings from the In hair cell and the ipsilateral type B photoreceptor. Recordings in the upper half of the figure are continued in the lower half. The hair cell (upper record in each half) impulse train (elicited by a +0.8-nA current step) caused a hyperpolarizing wave in the type B cell. Note that the hyperpolarizing wave at offset of the impulse train is followed by a long-lasting increase of type B impulse activity, whereas the hair cell remains hyperpolarized (i.e., more negative than its resting potential, indicated by the dashed line). Steady hyperpolarization of the type B cell throughout the same experiment revealed an increase of EPSPs after the hair cell impulse train. (From Tabata and Alkon, 1982)

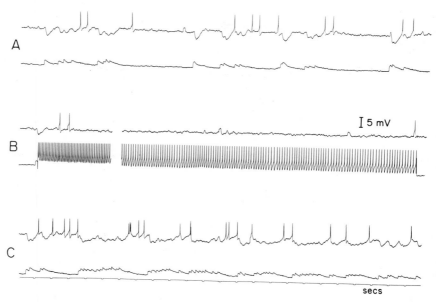

Figure 96. Simultaneous recordings from the caudal hair cell and the ipsilateral type B photoreceptor. Note the IPSPs recorded from the hair cell [upper records in (A), (B), and (C)] occur in synchrony with EPSPs recorded from the type B photoreceptor lower records. (A) Spontaneous synaptic potentials in darkness. (B); Type B impulse train (elicited by a +1.2-nA current step) causes cessation of IPSPs recorded from the hair cell. A record of ~ 8 s was deleted for illustration but showed no IPSPs. (C) The impulse train is followed by a marked increase in EPSP frequency (recorded from the B cell) and IPSP frequency (recorded from the hair cell). Records of (A), (B), and (C), except for the 8-s deletion (indicated by space), were continuous. (From Tabata and Alkon, 1982)

followed by a prolonged postinhibitory rebound depolarization of the E cell. Other optic ganglion cells are known to receive synaptic inhibition, and at least in one case, synaptic excitation from ipsilateral hair cells.

Interaction of Hermissenda visual, statocyst, and Chemosensory Pathways

Chemosensory stimuli delivered to the *Hermissenda* tentacles cause, via the tentacular nerve, synaptic excitation and, in other cases, inhibition of cerebropleural central neurons. Tentacular stimulation also causes synaptic inhibition of type B photoreceptors and hair cells.

Bibliography

Akaike, T., and Alkon, D. L. (1980). Sensory convergence on central visual neurons in *Hermissenda*. *J. Neurophysiol.* 44:501–13.

Alkon, D. L. (1973). Neural organization of a molluscan visual system. *J. Gen. Physiol.* 61:444–61.

Alkon, D. L. (1976). The economy of photoreceptor function in a primitive nervous system. In *Neural Principles in Vision*, ed. by F. Zettler and R. Weiler, pp. 410–26. Springer-Verlag, New York.

Alkon, D. L. (1983). Learning in a marine snail. *Sci. Am.* 249:70–84.

Alkon, D. L., and Fuortes, M. G. F. (1972). Responses of photoreceptors in *Hermissenda*. *J. Gen. Physiol.* 60:631–49.

Alkon, D. L., and Grossman, Y. (1978). Long-lasting depolarization and hyperpolarization in eye of *Hermissenda*. *J. Neurophysiol.* 41:1328–42.

Detwiler, P. B., and Alkon, D. L. (1973). Hair cell interactions in the statocyst of *Hermissenda*. *J. Gen. Physiol.* 62:618–42.

Goh, Y., and Alkon, D. L. (1984). Sensory, interneuronal and motor interactions within the *Hermissenda* visual pathway. *J. Neurophysiol.* 52:156–69.

Grossman, Y., Alkon, D. L., and Heldman, E. (1979). A common origin of voltage noise and generator potentials in statocyst hair cells. *J. Gen. Physiol.* 73:23–48.

Jerussi, T. P., and Alkon, D. L. (1981). Ocular and extraocular responses of identifiable neurons in pedal ganglia of *Hermissenda crassicornis*. *J. Neurophysiol.* 46:659–71.

Tabata, M., and Alkon, D. L. (1982). Positive synaptic feedback in the visual system of the nudibranch mollusc *Hermissenda crassicornis*. *J. Neurophysiol.* 48:174–91.

Index